Navier-Stokes Equations

Chicago Lectures in Mathematics Series
J. Peter May, Robert J. Zimmer, and Spencer J. Bloch, Editors

Other *Chicago Lectures in Mathematics* titles
available from The University of Chicago Press:

The Theory of Sheaves, by Richard G. Swan (1964)
Topics in Ring Theory, by I. N. Herstein (1969)
Fields and Rings, by Irving Kaplansky (1969; 2d ed. 1972)
Infinite Abelian Group Theory, by Phillip A. Griffith (1970)
Topics in Operator Theory, by Richard Beals (1971)
Lie Algebras and Locally Compact Groups, by Irving Kaplansky (1971)
Several Complex Variables, by Raghavan Narasimhan (1971)
Torsion-Free Modules, by Eben Matlis (1973)
The Theory of Bernoulli Shifts, by Paul C. Shields (1973)
Stable Homotopy and Generalized Homology, by J. F. Adams (1974)
Commutative Rings, by Irving Kaplansky (1974)
Banach Algebras, by Richard Mosak (1975)
Rings with Involution, by I. N. Herstein (1976)
Theory of Unitary Group Representation, by George W. Mackey (1976)
Infinite-Dimensional Optimization and Convexity, by Ivar Ekeland and Thomas Turnbull (1983)
Commutative Semigroup Rings, by Robert Gilmer (1984)

Peter Constantin and Ciprian Foias

Navier-Stokes Equations

The University of Chicago Press
Chicago and London

PETER CONSTANTIN is professor of mathematics at the University of Chicago.
CIPRIAN FOIAS is Distinguished Professor of Mathematics at Indiana University.

The University of Chicago Press, Chicago 60637
The University of Chicago Press, Ltd., London

Printed in the United States of America
97 96 95 94 93 92 91 90 89 5432

Library of Congress Cataloging-in-Publication Data

Constantin, P. (Peter), 1951–
Navier-Stokes equations / Peter Constantin and Ciprian Foias.
p. cm.—(Chicago lectures in mathematics)
Bibliography: p.
Includes index.
1. Navier-Stokes equations. I. Foias, Ciprian. II. Title.
III. Series.
QA374.C655 1989
515.3'53—dc19 88-13122
ISBN 0-226-11548-8. ISBN 0-226-11549-6 (pbk.) CIP

CONTENTS

INTRODUCTION

These constitute lecture notes of graduate courses given by the authors at Indiana University (1985-86) and the University of Chicago (1986-87), respectively.

In recent years there has been considerable progress in some of the questions related to the Navier-Stokes equations and their relation to finite-dimensional phenomena. For instance, the upper bound for the dimension of the universal attractor for 2D Navier-Stokes equations has been lowered from an estimate of the type $G^2 \exp G^4$ to an estimate of the type $G^{2/3} \log G^{1/3}$, where G is a nondimensional number, typically of the order of 100-1000. This most recent estimate can be understood in terms of the Kraichnan length and seems to be optimal for general body forces.

We try in these lecture notes to give an almost self-contained treatment of the topics we discuss. These notes are by no means an exhaustive treatise on the subject of Navier-Stokes equations. It has been our choice to present results using the most elementary techniques available. Thus, for instance, the regularity theory for the Stokes system (Chapter 3) is an adaptation of the classical L^2 regularity theory for a single elliptic equation of [A1]; our adaptation is inspired from [G]. Another example of our desire to illustrate the general results, while avoiding excessive technicalities, is the way we describe the asymptotic behavior of the eigenvalues of the Stokes operator (Chapter 4). For general bounded domains we provide a lower bound, using essentially elementary means. The lower bounds are all we really need later on. For completeness we give the elementary proof of the exact asymptotic behavior in the periodic case. The same asymptotic behavior for general domains, while true, would have required considerably more effort to describe. Questions regarding the notions of weak and strong solutions and their relations to classical solutions are studied in some detail. We prove that strong solutions are as smooth as the data permit; thus, loss of regularity can only occur if the solution ceases to be strong. We then show how, if there is an initial datum leading to loss of regularity in infinite time, there exists another one which leads to loss of regularity in finite time. We give the argument of Scheffer and Leray estimating the Hausdorff dimension of the singular times of a weak solution to be not more than 1/2. A simple argument is used to prove that, in the absence of boundaries, the vanishing viscosity limit of the Navier-Stokes equation is the Euler equation for incompressible fluids. The same technique can be used to show that as long as the solution to the incompressible Euler equation is smooth, solutions to small viscosity Navier-Stokes equations with the same initial data remain smooth. We provide a proof of time analyticity and consequent backward uniqueness for the initial value problem for the Navier-Stokes equations.

The importance of contact element transport is emphasized in the last chapter. We study first (Chapter 13) the decay of volume elements and give optimal lower bounds for the dimension at which this process starts. These bounds use inequalities of Lieb-Thirring. The construction of the universal attractor for 2D Navier-Stokes is given in Chapter 14. The fractal and Hausdorff dimensions of the universal attractor are estimated making the connection with the Kaplan-Yorke formula involving global Lyapunov exponents. Upper bounds for the fractal dimension of bounded invariant sets for 3D Navier-Stokes are given also. The final chapter deals with the concept of inertial manifolds for an artificial viscosity perturbation of the Navier-Stokes equation. The spectral blocking property and consequent cone invariance are illustrated in detail. These are ideas of independent interest and were successfully used to construct inertial manifolds for several physically significant equations. As of this writing the question of the existence of inertial manifolds for the Navier-Stokes equations remains open.

We wish to thank E. Titi, who taught some of the classes at both Indiana and Chicago and assisted in the preparation of these notes. We are indebted to Fred Flowers for his expert typing. This work was performed while PC was a Sloan research fellow.

1

NOTATION AND PRELIMINARY MATERIAL

Let $\Omega \subseteq \mathbb{R}^n$ be an open set. Ω is said to have the segment property if the boundary of Ω, $\partial\Omega$, has a locally finite open cover (U_i), $i \in I$ and for each i there exists a direction $\omega_i \in S^{n-1}$ and $\varepsilon_i > 0$ such that, for $x \in U_i \cap \overline{\Omega}$, $x_t = x + t\omega_i \in \Omega$ for $0 < t < \varepsilon_i$.

We denote by $L^p(\Omega) = \{f | f:\Omega \to \mathbb{R}, \text{ measurable}, \int|f(x)|^p dx < \infty\}$. We shall use $(\cdot\, , \cdot)$ for the scalar product in $L^2(\Omega)$. If Ω has the segment property the notions of a weak derivative in the sense of distributions and in the L^p sense coincide. We denote $\frac{\partial}{\partial x_i} = D_i$, $i = 1,\ldots,n$ and

$$D^\alpha = \frac{\partial^{|\alpha|}}{\partial x_1^{\alpha_1} \cdots \partial x_n^{\alpha_n}}, \quad |\alpha| = \alpha_1 + \cdots + \alpha_n.$$

$W^{m,p}(\Omega)$ are the Sobolev spaces $W^{m,p}(\Omega) = \{f | D^\alpha f \in L^p, |\alpha| \leq m\}$. For $p = 2$ we denote $H^m(\Omega) = W^{m,2}(\Omega)$. The norm in $W^{m,p}(\Omega)$ is

$$\|u\|_{m,p,\Omega} = \Big(\sum_{|\alpha| \leq m} \|D^\alpha u\|^2_{L^p(\Omega)}\Big)^{1/2}.$$

When $p = 2$ we write $\|u\|_{m,\Omega}$ instead of $\|u\|_{m,2,\Omega}$. $H^m(\Omega)$ is a Hilbert space equipped with the scalar product

$$(u,v)_{m,\Omega} = \sum_{|\alpha| \leq m} \int_\Omega (D^\alpha u)(x)(D^\alpha v)(x)dx.$$

The spaces $W^{m,p}(\Omega)$ are Banach spaces. We shall use the same notation $L^p(\Omega)$, $H^m(\Omega)$, $W^{m,p}(\Omega)$ for vectorial counterparts. For instance, the scalar product in $(H^m(\Omega))^n$ will be denoted

$$(\ , \ )_{m,\Omega}: \quad (u,v)_{m,\Omega} = \sum_{|\alpha| \leq m} \int_\Omega D^\alpha u \cdot D^\alpha v \, dx,$$

where . signifies scalar product in $\mathbb{R}^n$. We shall use sometimes the notation $< \ , \ >$ for the scalar product in $\mathbb{R}^n$. The closure of $C_0^\infty(\Omega)$ in $W^{m,p}(\Omega)$ is denoted by $W_0^{m,p}(\Omega)$.

Proposition 1.1. Let Ω satisfy the segment property. Then $C_0^\infty(\mathbb{R}^n)$ is dense in $W^{m,p}(\Omega)$, for $1 \leq p < \infty$.

The idea of the proof is the following. Let $u \in W^{m,p}(\Omega)$. We first approximate u in $W^{m,p}(\Omega)$ by a sequence of elements in $W^{m,p}(\Omega)$ with compact support by considering $u_m(x) = \phi(\frac{x}{m})u(x)$, where $\phi(x) = 1$ if $|x| \leq 1$, $\phi \in C_0^\infty(\mathbb{R}^n)$, supp $\phi \subset \{x \mid |x| \leq 2\}$. Then, using a partition of unity we may assume that the support of u is compact and either is included in Ω or is one of the sets U_i from the definition of the segment property. If the support of u is contained in Ω then a standard convolution with a mollifier will provide the approximation.

We may assume that the support of u is compact and included in some open set $V_i \subset\subset U_i$. Let $\tilde{u}$ be the extension of u defined by setting $\tilde{u}$ to be zero outside $\bar{\Omega}$. Then $\tilde{u} \in W^{m,p}(\mathbb{R}^n \setminus (\partial\Omega \cap V_i))$. We approximate $\tilde{u}$

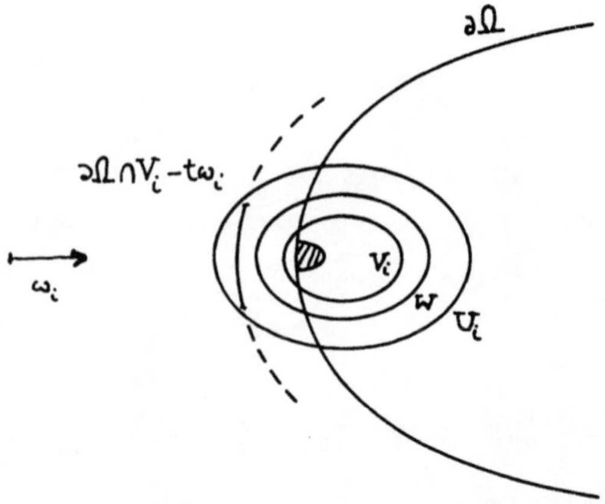

by $u_t = \tilde{u}(\cdot + t\omega_i)$ for small t. By doing this we push the singular set $\partial\Omega \cap V_i$ to $\partial\Omega \cap V_i - t\omega_i$: $u_t \in W^{m,p}(\mathbb{R}\backslash(\partial\Omega \cap V_i - t\omega_i))$. From the segment property this set does not touch $\bar{\Omega}$. Thus $u_t \in W^{m,p}(W)$ for some open neighborhood W of $\bar{\Omega} \cap V_i$. A convolution of u_t with some mollifier will produce a $C_0^\infty(\bar{\Omega})$ function near to u.

Proposition 1.2. (The Poincaré inequality) If Ω is bounded in some direction (i.e., if there exists a straight line in $\mathbb{R}^n$ such that the projection of Ω on it is bounded) then

$$(1.1) \qquad \|u\|_{L^2(\Omega)} \leq C(\Omega)\|\nabla u\|_{L^2(\Omega)} \qquad \text{for all } u \in H_0^1(\Omega).$$

We shall denote by $|u| = \|u\|_{L^2(\Omega)}$. Also, the Dirichlet norm $\|\nabla u\|_{L^2(\Omega)} = (\int_\Omega \sum_{i=1}^n |D_i u|^2 dx)^{1/2}$ will be denoted by $\|u\|$. This notation ($|u|$, $\|u\|$) will be used for vector valued functions, too, throughout this work.

Proof. Since $C_0^\infty(\Omega)$ is dense in $H_0^1(\Omega)$ it is enough to prove (1.1) for $\varphi \in C_0^\infty(\Omega)$. Performing a rotation if necessary we may assume, without loss of generality, that Ω is bounded in the x_1 direction. Let d be the width of Ω in the x_1 direction: $|x_1 - \tilde{x}_1| \leq d$ for all $x, \tilde{x}$ in Ω. Since the support of φ is compact in Ω there exists x_1^0 belonging to the projection of Ω on the x_1 axis, such that supp $\varphi \subset \{x \mid x_1 > x_1^0\}$. We consider φ to be defined in $\mathbb{R}^n$. Then

$$\varphi(x_1,x') = \varphi(x_1^0,x') + \int_{x_1^0}^{x_1} \frac{\partial\varphi}{\partial x_1}(s,x')ds = \int_{x_1^0}^{x_1} \frac{\partial\varphi}{\partial x_1}(s,x')ds.$$

Using the Cauchy-Schwartz inequality,

$$|\varphi(x_1,x')|^2 \leq |x_1 - x_1^0| \int_{x_1^0}^{x_1} |\frac{\partial\varphi}{\partial x_1}(s,x')|^2 dx' \leq d \int_{x_1^0}^{x_1} |\frac{\partial\varphi}{\partial x_1}(s,x')|^2 ds.$$

Integrating in x',

$$\int |\varphi(x_1,x')|^2 dx' \leq d\|\varphi\|^2.$$

Integrating in x_1 over the projection of the support of

$$|\varphi|^2 \leq d^2\|\varphi\|^2.$$

(The length of the projection of Ω on the x_1 axis is less than d.) We see that the constant $C(\Omega)$ in (1.1) can be taken to be d, the width of Ω in some direction.

We shall present now some trace theorems and the Weyl decomposition from [T1]. For more details see [T1].

Let Ω have the segment property. Let $E(\Omega)$ denote the space $E(\Omega) = \{u \in L^2(\Omega)^n \mid \text{div } u \in L^2(\Omega)\}$. We denoted $\text{div } u = \nabla \cdot u = \sum_{i=1}^{n} \frac{\partial u_i}{\partial x_i}$, the derivatives taken in the sense of distributions in Ω. $E(\Omega)$ is a Hilbert space with the scalar product

$$[u,v] = (u,v) + \int_\Omega (\text{div } u)(\text{div } v)dx.$$

<u>Proposition 1.3</u>. The set $(C_0^\infty(\mathbb{R}^n))^n$ is dense in $E(\Omega)$.

<u>Proof</u>. Same method as for Proposition 1.1.

We shall impose a much more restrictive assumption on Ω. We shall say that Ω is of class C^r if there exists a locally finite open cover $(U_i)_{i \in I}$ of $\partial\Omega$ and C^r diffeomorphisms $\psi_i : U_i \to D$, where D is the unit open disk in $\mathbb{R}^n$, $D = \{x \mid |x| < 1\}$ such that

$$\psi_i(U_i \cap \Omega) = D_+ = \{x \in D \mid x_n > 0\},$$

$$\psi_i(U_i \cap \partial\Omega) = D_0 = \{x \in D \mid x_n = 0\}.$$

Suppose now that Ω is bounded and of class C^2. The trace operator $\gamma_0: H^1(\Omega) \to L^2(\partial\Omega)$ is a bounded linear operator agreeing with the restriction operation $u \longmapsto u/\partial\Omega$ for continuously differentiable functions on $\overline{\Omega}$. The kernel of γ_0 is $H_0^1(\Omega)$. The image is denoted $H^{1/2}(\partial\Omega)$, and is a Hilbert space. (On $\partial\Omega$ we consider the Lebesgue measure.) There exists a lifting operator $\ell_\Omega: H^{1/2}(\partial\Omega) \to H^1(\Omega)$ which satisfies $\gamma_0 \ell_\Omega$ = identity in $H^{1/2}(\partial\Omega)$. For these results see [Li1], [L-M1]. We define $H^{-1/2}(\partial\Omega)$ as the dual space of $H^{1/2}(\partial\Omega)$. We want to define the normal component $u \cdot n_\Omega$ of elements of $E(\Omega)$. The notation n_Ω stands for the outer normal to $\partial\Omega$.

Proposition 1.4. Let Ω be an open bounded set of class C^2. There exists a continuous linear operator $\gamma: E(\Omega) \to H^{-1/2}(\Omega)$ such that $\gamma(u) = u \cdot n_\Omega$ for every $u \in C_0^\infty(\overline{\Omega})^n$. The Stokes formula

(1.2) $$(u, \text{grad } w) + (\text{div } u, w) = \langle \gamma(u), \gamma_0(w) \rangle$$

holds for every $u \in E(\Omega)$, $w \in H^1(\Omega)$.

The idea of the proof is to use the lifting operator $\ell_\Omega: H^{1/2}(\partial\Omega) \to H^1(\Omega)$ to define the element $\gamma(u)$ of the dual $H^{-1/2}(\partial\Omega)$ of $H^{1/2}(\Omega)$ by (1.2):

$$\langle \gamma(u), \phi \rangle \stackrel{\text{def}}{=} (u, \text{grad } \ell_\Omega \phi) + (\text{div } u, \ell_\Omega \phi)$$

for all $\phi \in H^{1/2}(\partial\Omega)$, and fixed $u \in E(\Omega)$. Clearly, since $\ell_\Omega \phi \in H^1(\Omega)$ and $\|\ell_\Omega(\phi)\|_{H^1(\Omega)} \le c\|\phi\|_{H^{1/2}(\partial\Omega)}$, $|\langle \gamma(u), \phi \rangle| \le c\|\phi\|_{H^{1/2}(\partial\Omega)} \|u\|_{E(\Omega)}$. This takes care of the fact that $\gamma: E(\Omega) \to H^{-1/2}(\partial\Omega)$ and γ is a bounded linear map. If u is a $C^\infty(\overline{\Omega})^n$ function and ϕ is the restriction to $\partial\Omega$ of a $C^\infty(\overline{\Omega})$ function, w, the divergence theorem (Stokes formula) implies that

$$\int_{\partial\Omega} (u \cdot n_\Omega)\phi \, dx = (u, \text{grad } w) + (\text{div } u, w).$$

Since $w - \ell_\Omega(\phi)$ is in the kernel of γ_0, that is, $w_0 = w - \ell_\Omega(\phi) \in H_0^1(\Omega)$ and since $(u, \text{grad } w_0) + (\text{div } u, w_0) = 0$ for any $w_0 \in H_0^1(\Omega)$ it follows that $\int_{\partial\Omega} (u \cdot n_\Omega)\phi \, dx = \langle\gamma(u), \phi\rangle$. Now the functions ϕ which are restrictions of $C^\infty(\overline{\Omega})$ functions are dense in $H^{1/2}(\partial\Omega)$. It follows that $u \cdot n_\Omega = \gamma(u)$ for smooth u.

<u>Proposition 1.5</u>. Let $u \in E(\Omega)$ be such that $\gamma(u) = 0$. Then u can be approximated in $E(\Omega)$ by functions belonging to $C_0^\infty(\Omega)^n$.

For the proof we note that $\gamma(u) = 0$ implies $\int_\Omega (\text{div } u)\varphi + u \cdot \text{grad } \varphi \, dx = 0$ for every $\varphi = \Phi|_{\overline{\Omega}}$, $\Phi \in C_0^\infty(\mathbb{R}^n)$. We define by $\tilde{u}$ the extension by 0 of u: $\langle \tilde{u}_i, \psi\rangle = \int_\Omega u_i(x)\psi(x)dx$, $i = 1, \ldots, n$, $\psi \in C_0^\infty(\mathbb{R}^n)$. It follows that $\text{div } \tilde{u} = \widetilde{\text{div } u}$. Thus $\tilde{u} \in E(\mathbb{R}^n)$. Now we localize, as in the proof of Proposition 1.1. We may assume that the support of u is compact and contained in one of the sets U_i from the definition of the segment property. However, in the present case, there is no singular set since $\tilde{u} \in E(\mathbb{R}^n)$ and thus we can perform a small translation "inland", $u_t = \tilde{u}(\cdot - t\omega_i)$ (as opposed to $u_t = \tilde{u}(\cdot + t\omega_i)$ in the proof of Proposition 1.1). By this translation we detach the support of u from the boundary: $u_t \in E(\mathbb{R}^n)$, $\text{supp } u_t \subset \Omega$. Again convolution with a mollifier will produce functions in $C_0^\infty(\mathbb{R}^n)$ close to u but this time their support will be included in Ω.

Let us denote by $\mathcal{V}$ the set

$$(1.3) \qquad \mathcal{V} = \{\varphi \in (C_0^\infty(\Omega))^n \mid \text{div } \varphi = 0\}.$$

Let us denote by H and V the closure of $\mathcal{V}$ in $L^2(\Omega)^n$ and $H_0^1(\Omega)^n$, respectively.

$$(1.4) \qquad H = \text{closure of } \mathcal{V} \text{ in } L^2(\Omega)^n$$

$$(1.5) \qquad V = \text{closure of } \mathcal{V} \text{ in } H_0^1(\Omega)^n.$$

Let $\Omega \subset \mathbb{R}^n$ be open. We state without proof the following results.

Proposition 1.6. Let $f_i \in \mathcal{D}'(\Omega)$, $i = 1,\ldots,n$, be distributions. Then $f = \text{grad } p$ for some $p \in \mathcal{D}'(\Omega)$ if and only if $\langle f,v\rangle = 0$ for all $v \in \mathcal{V}$.

Also

Proposition 1.7. Let $\Omega \subset \mathbb{R}^n$ be an open bounded set with locally Lipschitz boundary.

(i) If a distribution $p \in \mathcal{D}'(\Omega)$ has all its first derivatives $D_i p$ in $L^2(\Omega)$ then $p \in L^2(\Omega)$ and

$$\|p\|_{L^2(\Omega)/\mathbb{R}} \le c(\Omega)\|\nabla p\|_{L^2(\Omega)^n}$$

(ii) If a distribution p has all its first derivatives in $H^{-1}(\Omega)$ (the dual of $H^1(\Omega)$) then $p \in L^2(\Omega)$ and

$$\|p\|_{L^2(\Omega)/\mathbb{R}} \le c(\Omega)\|\nabla p\|_{H^{-1}(\Omega)^n}$$

In both cases, if no restriction is imposed on $\partial\Omega$ it follows that $p \in L^2_{loc}(\Omega)$. By $\|p\|_{L^2(\Omega)/\mathbb{R}}$ we mean

$$\inf_{c \in \mathbb{R}} \|p - c\|_{L^2(\Omega)} = \left\| p - \frac{\int_\Omega p\,dx}{|\Omega|} \cdot 1 \right\|_{L^2(\Omega)}$$

Proposition 1.8. Let $\Omega \subset \mathbb{R}^n$ be a locally Lipschitz bounded open set. Then

(1.6) $$H = \{u \in L^2(\Omega)^n \mid \text{div } u = 0,\ \gamma(u) = 0\}$$

(1.7) $$H^\perp = \{u \in L^2(\Omega) \mid u = \text{grad } p,\ p \in H^1(\Omega)\}.$$

Proof. For (1.7): If $u = \text{grad } p$ with $p \in H^1(\Omega)$ then $\langle u,v\rangle = 0$ for all $v \in \mathcal{V}$ and $u \in H^\perp$. On the other hand, $\langle u,v\rangle = 0$ for all $v \in \mathcal{V}$ implies by Propositions 1.6 and 1.7, $u = \text{grad } p$, $p \in H^1(\Omega)$.

Now for the proof of (1.6). Denote by $H^{\bullet}$ the right hand side of (1.6). If u belongs to H then u is the limit in $L^2(\Omega)^n$ of a sequence of functions in $\mathcal{V}$. Thus clearly div u = 0. Therefore $u \in E(\Omega)$ and the convergence of the functions of $\mathcal{V}$ to u takes place in $E(\Omega)$. Now $\gamma: E(\Omega) \to H^{-1/2}(\partial\Omega)$ is continuous. Therefore, $\gamma(u) = 0$. Now $H \subset H^{\bullet}$. Moreover H is dense in the $L^2(\Omega)^n$ topology in $H^{\bullet}$. For $H^{\bullet}$ is a closed subspace of $L^2(\Omega)^n$ and if $H^{\bullet} \ominus H$ would be nonempty, say $v \in H^{\bullet} \ominus H$, then $v \in H^{\perp}$, and thus $v = \nabla p$ with $p \in H^1(\Omega)$ and also $v \in H^{\bullet}$, thus div (grad p) = $\Delta p = 0$, $\gamma(u) = \frac{\partial p}{\partial n_\Omega} = 0$. Thus p must be constant on each connected component of Ω. Thus u = 0, since H is closed $H = H^{\bullet}$.

Proposition 1.9. Let Ω be open, bounded, connected of class C^2. Then $L^2(\Omega) = H \oplus H_1 \oplus H_2$, where H, H_1, H_2 are mutually orthogonal spaces,

$$H_1 = \{u \in L^2(\Omega)^n \mid u = \text{grad } p,\ p \in H^1(\Omega),\ \Delta p = 0\}$$

and

$$H_2 = \{u \in L^2(\Omega)^n \mid u = \text{grad } p,\ p \in H_0^1(\Omega)\}.$$

Proof. Clearly H_1, H_2 are included in $H^{\perp}$. Also H_1 and H_2 are orthogonal, for if $u = \nabla p$, $v = \nabla q$, $p \in H^1(\Omega)$, $\Delta p = 0$ and $q \in H_0^1(\Omega)$ then, by the Stokes formula (1.2)

$$(u,v) = (u,\nabla q) = (\gamma u, \gamma_0 q) - (\Delta p, q) = 0.$$

Let now $u \in L^2(\Omega)^n$ be arbitrary. First let us solve $\Delta p = \text{div } u \in H^{-1}(\Omega)$, $p \in H_0^1(\Omega)$. This Dirichlet problem has a unique solution. We set $u_2 = \nabla p$; clearly $u_2 \in H_2$. Then, for $u - u_2$ we solve the Neumann problem $\Delta q = 0$, $\frac{\partial q}{\partial n_\Omega} = \gamma(u - u_2)$. Remark first that $\text{div}(u - u_2) = \text{div } u - \Delta p = 0$ and so $u - u_2 \in E(\Omega)$. Moreover, the compatibility condition $\langle \gamma(u - u_2), 1 \rangle = 0$ is satisfied because of the Stokes formula. Then there exists a unique q (up to additive constants) such that $q \in H^1(\Omega)$ and solves the Neumann problem. Then $u_1 = \nabla q$ clearly is in H_1.

Finally, let $u_0 = u - u_1 - u_2$. We need to show that $u_0 \in H$. Clearly,

$$\begin{aligned} \operatorname{div} u_0 &= \operatorname{div} u - \operatorname{div} u_1 - \operatorname{div} u_2 = \\ &= \operatorname{div} u - \Delta q - \Delta p = \operatorname{div} u - 0 - \operatorname{div} u = 0. \end{aligned}$$

Also,

$$\gamma(u_0) = \gamma(u - u_2) - \gamma(u_1) = \gamma(u - u_2) - \frac{\partial q}{\partial n_\Omega} = 0.$$

<u>Remark 1.10</u>. Let $P: L^2(\Omega)^n \to H$ be the orthogonal projector $u \mapsto u_0$. We shall refer to it occasionally as the Leray Projector. If $u \in H_0^1(\Omega)^n$ then $Pu \in H^1(\Omega)^n$. Indeed, in finding u_2 we solve now $\Delta p = \operatorname{div} u \in L^2(\Omega)$, $p \in H_0^1(\Omega)$ so we can find $p \in H^2(\Omega)$. Therefore $u_2 = \nabla p$ belongs to $H^1(\Omega)$. Then $u - u_2$ belongs to $H^1(\Omega)$ and $\gamma(u - u_2)$ belongs to $H^{1/2}(\partial\Omega)$. Thus solving the Neumann problem $\Delta q = 0$, $\frac{\partial q}{\partial n_\Omega} = \gamma(u - u_2)$ we obtain $q \in H^2(\Omega)$ and thus $u_1 = \nabla q$ belongs to H^1. It follows that $u_0 = u - u_1 - u_2$ belongs to $H^1(\Omega)$. Thus $P: (H_0^1(\Omega))^n \to (H^1(\Omega))^n$ is bounded.

If further smoothness is assumed on $\partial\Omega$ then we see that P is bounded in higher Sobolev spaces.

<u>Proposition 1.11</u>. Let Ω be open, bounded and locally Lipschitz. Then

$$V = \{u \in H_0^1(\Omega)^n \mid \operatorname{div} u = 0\} \tag{1.8}$$

<u>Proof</u>. Let $V^\cdot$ be the space defined in the right hand side of (1.8). Clearly $V \subset V^\cdot$. Moreover $V^\cdot$ is closed in $H_0^1(\Omega)^n$. Let L be any continuous linear functional on $V^\cdot$. Clearly, L can be extended to $H^1(\Omega)^n$ so it can be written as $L = \sum_{i=1}^{n} (\ell_i, \cdot)$ with $\ell_i \in H^{-1}(\Omega)$. Now, if $\langle L, v\rangle = 0$ for all $v \in V$ it follows from Proposition 1.6 that there

exists $p \in \mathcal{D}'(\Omega)$ such that $\ell_i = \frac{\partial p}{\partial x_i}$. Since $\ell_i \in H^{-1}(\Omega)$ it follows from Proposition 1.7 that $p \in L^2(\Omega)$. But then

$$L(w) = \sum_{i=1}^{n} \langle \ell_i, w_i \rangle = \sum_{i=1}^{n} \langle D_i p, w_i \rangle = -\int_\Omega p \operatorname{div} w = 0$$

for every $w \in V^{\bullet}$. Thus V is dense in $V^{\bullet}$ and since V is closed, $V = V^{\bullet}$.

2

THE STOKES EQUATIONS. EXISTENCE AND UNIQUENESS OF WEAK SOLUTIONS

Let Ω be an open bounded set in $\mathbb{R}^n$. Let $f \in L^2(\Omega)^n$. The Stokes equations for the vector $u = (u_1,\ldots,u_n)$ and the scalar f are ($\nu > 0$ is a constant)

(2.1) $\qquad -\nu\Delta u + \operatorname{grad} p = f \quad \text{in } \Omega$

(2.2) $\qquad \operatorname{div} u = 0 \quad \text{in } \Omega$

(2.3) $\qquad u = 0 \quad \text{on } \partial\Omega.$

If u,p are smooth then integating by parts we obtain

(2.4) $\qquad \nu((u,v)) = (f,v)$

for all $v \in \mathcal{V}$. Hereafter ((u,v)) is the scalar product
$((u,v)) = \sum_{i=1}^{n} (D_i u, D_i v)$.

We shall say that u is a weak solution of the Stokes equations (2.1)-(2.3) if

(2.5) $$\begin{cases} u \in V \text{ and} \\ \nu((u,v)) = (f,v) \quad \text{for all } v \in \mathcal{V}. \end{cases}$$

Let us note that (2.5) implies by continuity that $\nu((u,v)) = (f,v)$ for all $v \in V$.

<u>Proposition 2.1</u>. Let Ω be open bounded and of class C^2. The following are equivalent

(i) u is a weak solution of the Stokes equations

(ii) $u \in H_0^1(\Omega)^n$ and satisfies: there exists $p \in L^2(\Omega)$ such that

(2.6) $-\nu\Delta u + \text{grad } p = f \quad \text{in} \quad \mathcal{D}'(\Omega)$

(2.7) $\text{div } u = 0 \qquad \text{in} \quad \mathcal{D}'(\Omega)$

(2.8) $\gamma_0(u_i) = 0 \qquad i = 1,\ldots,n.$

(iii) $u \in V$ achieves the minimum of $\Phi(v) = \nu\|v\|^2 - 2(f,v)$ on V.

Proof. If (ii) is true then $u \in V$ because of (2.7) and (1.8). Then (2.4) is true because of (2.6). So (ii) implies (i). Conversely, if (i) is true then $-\nu\Delta u - f$ is a distribution in $H^{-1}(\Omega)^n$ satisfying $\langle -\nu\Delta u - f, v\rangle = 0$ for all $v \in \mathcal{V}$. Then, by Propositions 1.6 and 1.7, $-\nu\Delta u - f$ is the gradient of an $L^2(\Omega)$ function. Now if (i) is true then $\Phi(u + w) = \nu\|u + w\|^2 - 2(u + w,f) = \Phi(u) + \nu\|w\|^2 \geq \Phi(u)$ for all $w \in V$. Conversely, if (iii) is satisfied then the expression $\Phi(u + \lambda v) - \Phi(u)$ is nonnegative for all $v \in V$, $\lambda \in \mathbb{R}$. But this expression is the quadratic polynomial $\lambda^2\nu^2\|v\|^2 + 2\lambda[\nu((u,v)) - (f,v)]$ and therefore the coefficient of λ must vanish.

Proposition 2.2 (Lax-Milgram). Let X be a separable Hilbert space and $a: X \times X \to R$ be a bilinear continuous coercive form. That is, if $\|\cdot\|_X$ denotes the norm in X

(i) $|a(u,v)| \leq c\|u\|_X\|v\|_X$

(ii) $a(u,u) \geq \alpha\|u\|_X^2$.

Then, for each ℓ linear continuous form on X, there exists a unique element $u_\ell \in X$ such that

(iii) $a(u_\ell,v) = \langle \ell,v\rangle$ for all $v \in X$

Proof. $a(\cdot,\cdot)$ is a scalar product in X. It induces a norm which is equivalent to the original norm. Then ℓ is a linear continuous form on X with this scalar product. By the F. Riesz representation theorem there exists and is unique u_ℓ such that (iii) is true.

Theorem 2.3. Let Ω be open and bounded in some direction. Then for every $f \in L^2(\Omega)^n$, $\nu > 0$ there exists a unique weak solution of the Stokes equations (2.1)-(2.3).

Proof. By the Poincaré inequality $a((\cdot,\cdot))$ is coercive on V. The result follows from the Lax-Milgram theorem. A second proof can be given using the characterization (iii) of Proposition 2.2.

3

REGULARITY OF SOLUTIONS OF THE STOKES EQUATIONS

The Stokes system is elliptic in the sense of Agmon, Douglis, and Nirenberg and therefore the general a priori estimates of [ADN 1,2] are available for it. The L^p existence and regularity results were first obtained by Vorovich and Yudovitch [V], Cattabriga [Ca1] and Solonnikov [Sol]. We shall restrict ourselves to the L^2 results. In the case of L^2 elliptic estimates the classical method of [A1] used to derive regularity results for scalar equations can be adapted with no difficulty to the Stokes system. We use thus the notation and technique of [A1].

Definition 3.1. We define the difference operators δ_h^i by

$$(\delta_h^i u)(x) = \frac{1}{h}(u(x + he^i) - u(x)), \quad h \neq 0.$$

Here $e^i = (\delta_{i1}, \delta_{i2}, \ldots, \delta_{in})$ is the canonical basis of $\mathbb{R}^n$.

Lemma 3.2. Suppose $u \in H^m(\Omega)$, $m \geq 1$, $\overline{\Omega}' \subset \Omega$. Assume $\mathrm{dist}(\overline{\Omega}', \partial\Omega) > h > 0$. Then

$$(3.1) \qquad \| \delta_{\pm h}^i u \|_{m-1,\Omega'} \leq \|u\|_{m,\Omega}$$

Proof. For any function $f \in C^1(a, b+h)$ we have

$$f(x + h) - f(x) = \int_x^{x+h} f'(t)dt$$

and thus

$$|f(x+h) - f(x)|^2 \le h \int_x^{x+h} |f'(t)|^2 dt.$$

Integrating, we get

$$\int_a^b |f(x+h) - f(x)|^2 dx \le h \int_a^b dx \int_x^{x+h} |f'(t)|^2 dt$$

$$= h\Big[\int_a^{a+h} |f'(t)|^2 \Big(\int_a^t dx\Big)dt + \int_{a+h}^b |f'(t)|^2 \Big(\int_{t-h}^t dx\Big)dt + \int_b^{b+h} |f'(t)|^2 \Big(\int_{t-h}^b dx\Big)dt\Big]$$

$$\le h^2 \int_a^{b+h} |f'(t)|^2 dt.$$

Therefore

$$(3.2) \qquad \int_a^b \Big|\frac{f(x+h) - f(x)}{h}\Big|^2 dx \le \int_a^{b+h} |f'(t)|^2 dt.$$

Now clearly if $u \in C^m(\Omega)$ then using (3.2), the fact that δ_h^i commutes with differentiation and iterated integrals, we obtain (3.1). For general u the result follows from density arguments.

Lemma 3.3. Suppose Ω has the segment property. Assume $u \in H^m(\Omega)$ and that there exists a constant $C > 0$ such that, for every $\Omega' \subset\subset \Omega$, $\| \delta_h^i u \|_{m,\Omega'} \le C$ for all h sufficiently small. Then $\|D_i u\|_{m,\Omega} \le \gamma_m C$, where $\gamma_m = \sum_{|\alpha| = m} 1$.

Proof. Assume first $m = 0$. Fix $\Omega' \subset\subset \Omega$. By the weak compactness property of L^2 we find a sequence $\{h_k\}$ of reals $h_k \to 0$ and a function $u_i \in L^2$ such that $\delta_{h_k}^i u \xrightarrow[k]{} u_i$ weakly in $L^2(\Omega')$. Clearly also $\|u_i\|_{0,\Omega'} \le C$. For any $\varphi \in C_0^\infty(\Omega)$ we have

$$\int_{\Omega'} u_i \phi dx = \lim_k \int_\Omega (\delta_{h_k}^i u)\phi dx = - \lim_k \int_\Omega u\, \delta_{-h_k}^i \phi = - \int_{\Omega'} u D_i \phi dx.$$

This proves that $u_i = D_i u$ in the weak sense in Ω'. Allowing Ω' to vary we obtain the conclusion of the lemma for $m = 0$. For general m, the result follows from the $m = 0$ case applied to derivatives of order m of u.

Lemma 3.4. Let $R > 0$ and let $G_R = \{x \in \mathbb{R}^n \mid |x| < R, x_n > 0\}$. Suppose $u \in L^2(G)$ and assume that there exists a number C such that, for every $R' < R$, $\| \delta_h^i u \|_{0,G_{R'}} \leq C$ for some $i \in \{1,2,\ldots,n-1\}$. Then the weak derivative $D^i u$ belongs to $L^2(G_R)$ and

$$\|D^i u\|_{0,G_R} \leq C.$$

Proof. Same as before.

Remark 3.5. For vector valued functions, Lemmas 3.2, 3.3 and 3.4 are true. The difference operators act on each component. The operators δ_h^i act almost like derivatives:

$$\delta_h^i(av) = \delta_h^i(a)v + \tau_h^i(a)\delta_h^i(v) = a\delta_h^i(v) + \delta_h^i(a)\tau_h^i(v) \tag{3.3}$$

Here $\tau_h^i(a)(x) = a(x + he^i)$ is a translation operator and, of course, for small h, it is close to identity.

We are ready to give the main lemma. The method of proof of regularity involves localization (a partition of unity) and local flattening of the boundary. The change of variables by which the local flattening is obtained will transform the constant coefficients Stokes system into a slightly more complicated, variable coefficients system. The stragegy, thus, consists in treating systems of a type that is invariant to change of variables in simple domains. The operator $-\sum_{i,j=1}^{n} \frac{\partial}{\partial x_i}(a_{ij}(x)\frac{\partial}{\partial x_j})$ is said to be uniformly elliptic in a domain G if there exists $M > 0$ such that

(3.4) $$\frac{1}{M}|\xi|^2 \leq \sum_{i,j} a_{ij}(x)\xi_i\xi_j \leq M|\xi|^2, \quad \text{for all } x \in G,\ \xi \in \mathbb{R}^n.$$

From now on we shall use the summation convention (repeated indices are summed) unless the contrary is expressly stated. We shall consider two kinds of domains G, balls and half balls:

(3.5) $$G_R = \{x \in \mathbb{R}^n \mid |x| < R\}$$

(3.6) $$G_R = \{x \in \mathbb{R}^n \mid x_n > 0,\ |x| < R\}.$$

We shall denote by $\tilde{G}_R$ the set

(3.7) $$\tilde{G}_R = \{x \in \mathbb{R}^n \mid x_n \geq 0,\ |x| < R\}.$$

A function whose support is compact and included in $\tilde{G}_R$ may not vanish for points on $x_n = 0$.

<u>Lemma 3.6</u>. ([G]) Let $0 < R' < R$. Consider a weak solution v,p of the system

(3.8) $$-\frac{\partial}{\partial x_i}\left(a_{ij}(x)\frac{\partial v_m}{\partial x_j}\right) + b_j(x)\frac{\partial v_m}{\partial x_j} + g_{mj}(x)\frac{\partial p}{\partial x_j} = f_m, \quad m = 1,\dots,n$$

(3.9) $$g_{mk}(x)\frac{\partial v_m}{\partial x_k} = \rho(x)$$

where $$a_{ij} \in C^1(G_R), \quad g_{mj} \in C^2(G_R), \quad b_j \in C^0(G_R),$$

and $$f = (f_m)_{m=1,\dots,n} \in (L^2(G_R))^n, \quad \rho \in H^2(G_R).$$

The principal part of (3.8) is assumed to be uniformly elliptic, i.e., (3.4) holds. The domain G_R is either a ball ((3.5)) or a half ball ((3.6)).

Suppose $v \in (H_0^1(G_{R'}))^n$, $p \in L^2(G_{R'})$. Assume that the supports of v,p are compact in $G_{R'}$ ($\bar{G}_{R'}$ in case of the half balls). Then, there exists a constant C depending on R, R' and the coefficients of (3.8), (3.9) such that

$$(3.10) \quad \|D_i v\|_{H^1(G_R)} \le C[\|f\|_{L^2(G_R)} + \|v\|_{H^1(G_R)} + \|\rho\|_{H^2(G_R)} + \|p\|_{L^2(G_R)}]$$

where $i = 1,\dots,n$ if $G_R, G_{R'}$ are balls and $i = 1,2,\dots,n-1$ if G_R, $G_{R'}$ are half balls.

Proof. According to Lemmas 3.3 and 3.4 all we need to show is that the right hand side of (3.10) is an upper bound for $\| \delta_h^i v \|_{H^1(G_{R''})}$ for all $R' < R'' < R$ and $|h| < \frac{R - R''}{2}$. (The index i is fixed and not summed below.) The sense in which (3.8) is satisfied is the following: for every $\varphi \in (C_0^\infty(G_R))^n$ one has

$$(3.11) \quad \int a_{kj} \frac{\partial v_m}{\partial x_j} \frac{\partial \varphi_m}{\partial x_k} dx + \int b_j \frac{\partial v_m}{\partial x_j} \varphi_m dx - \int p \frac{\partial}{\partial x_j} (g_{mj}\varphi_m) dx = \int f_m \varphi_m.$$

Clearly (3.11) is true, by continuity, for every $\varphi \in (H_0^1(G_R))^n$. If $|h| < \frac{R - R''}{2}$ then if $\varphi \in C_0^\infty(G_{R''})^n$ we may apply (3.11) with φ replaced by $\varphi_h = -\delta_{-h}^i \varphi$. We obtain

$$(3.12) \quad \int \delta_h^i (a_{kj} \frac{\partial v_m}{\partial x_j}) \frac{\partial \varphi_m}{\partial x_k} dx - \int b_j \frac{\partial v_m}{\partial x_j} \delta_{-h}^i \varphi_m dx + \int p \frac{\partial}{\partial x_j} (g_{mj} \partial_{-h}^i \varphi_m)$$
$$= \int f_m \delta_{-h}^i \varphi_m.$$

We treat the three terms on the left-hand side of (3.12) separately. The first one using (3.3) can be computed

$$I = \int \delta_h^i \left(a_{kj} \frac{\partial v_m}{\partial x_j}\right) \frac{\partial \varphi_m}{\partial x_k} dx = \int a_{kj} \left(\delta_h^i \frac{\partial v_m}{\partial x_j}\right) \frac{\partial \varphi_m}{\partial x_k} dx + \int \delta_h^i (a_{kj}) \tau_h^i \frac{\partial v_m}{\partial x_j} \frac{\partial \varphi_m}{\partial x_k}$$

$$= a(\delta_h^i v, \varphi) + \int \delta_h^i (a_{kj}) \tau_h^i \left(\frac{\partial v_m}{\partial x_j}\right) \frac{\partial \varphi_m}{\partial x_k}$$

In view of the fact that a_{kj} are uniformly Lipschitz we obtain

$$|I - a(\delta_h^i v, \varphi)| \leq c\|v\|_{H^1(G_R)} \|\nabla \varphi\|_{L^2(G_{R''})}$$

The constant C is independent of h and will change during the proof.

$$(3.13) \qquad a(v, w) = \int a_{kj} \frac{\partial v_m}{\partial x_j} \frac{\partial w_m}{\partial x_k} dx.$$

We estimate the second term using Lemma 3.2:

$$|II| = \left|\int b_j \frac{\partial v_m}{\partial x_j} \delta_{-h}^i \varphi_m dx\right| \leq C\|v\|_{H^1(G_R)} \|\varphi\|_{H^1(G_{R'''})}$$

where $R''' = \dfrac{R + R''}{2}$.

In order to estimate the third term we write first, using (3.3)

$$\frac{\partial}{\partial x_j} (g_{mj} \delta_{-h}^i \varphi_m) = \delta_{-h}^i \left(g_{mj} \frac{\partial \varphi_m}{\partial x_j}\right) - \delta_{-h}^i (g_{mj}) \tau_{-h}^i \left(\frac{\partial \varphi_m}{\partial x_j}\right) + \frac{\partial g_{wj}}{\partial x_j} \delta_{-h}^i \varphi_m.$$

Now since supp p is compact in $G_{R'}$ $(\tilde{G}_{R'})$ we can find $\alpha \in C_0^2(G_{R''})$ such that $\alpha p = p$. Thus

$$\alpha(x) \frac{\partial}{\partial x_j} (g_{mj} \delta_{-h}^i \varphi_m) = \delta_{-h}^i \left(\alpha g_{mj} \frac{\partial \varphi_m}{\partial x_j}\right) - \delta_{-h}^i (\alpha) \tau_{-h}^i \left(g_{mj} \frac{\partial \varphi_m}{\partial x_j}\right)$$
$$- \alpha \delta_{-h}^i (g_{mj}) \tau_{-h}^i \left(\frac{\partial \varphi_m}{\partial x_j}\right) + \alpha \frac{\partial g_{mj}}{\partial x_j} \delta_{-h}^i \varphi_m$$

and therefore we obtain

$$|III| = \left|\int \alpha p \frac{\partial}{\partial x_j} (g_{mj} \delta_{-h}^i \varphi) dx\right|$$
$$\leq C\|p\|_{L^2(G_R)} \left[\left\| \delta_{-h}^i \left(\alpha g_{mj} \frac{\partial \varphi_m}{\partial x_j}\right) \right\|_{L^2(G_{R'''})} + \|\varphi\|_{H^1(G_{R'''})}\right.$$

The precaution of inserting α is not needed for the proof of the present lemma but it will take care of a minor fine point later on. The right-hand side of (3.12) is bounded by $\|f\|_{L^2(G_R)}\|\varphi\|_{H^1(G_{R'''})}$. Summing up, we obtained

$$(3.14)\quad |a(\delta_h^i v,\varphi)| \le C\{\|\varphi\|_{H^1(G_{R'''})}[\|f\|_{L^2(G_R)} + \|v\|_{H^1(G_R)} + \|p\|_{L^2(G_R)}] + \|p\|_{L^2(G_R)}\| \delta_{-h}^i(\alpha g_{mj}\frac{\partial\varphi_m}{\partial x_j}) \|_{L^2(G_{R'''})}\}$$

valid for every $\varphi \in (C_0^\infty(G_{R''}))^n$. Now for each fixed h, (small) $\delta_h^i v \in (H_0^1(G_{R''}))^n$. There exists a sequence $\varphi^{(\ell)} \in (C_0^\infty(G_{R''}))^n$ converging to $\delta_h^i v$ in $H^1(G_{R''})^n$. But then $\alpha g_{mj}\dfrac{\partial\varphi_m^{(\ell)}}{\partial x_j}$ converges in $L^2(G_{R''})$ to the corresponding expression $\alpha g_{mj}\delta_h^i\dfrac{\partial v_m}{\partial x_j}$. It follows that $\delta_{-h}^i(\alpha g_{mj}\dfrac{\partial \varphi_m^{(\ell)}}{\partial x_j})$ converges weakly in $L^2(G_{R'''})$ to $\delta_{-h}^i(\alpha g_{mj}\delta_h^i\dfrac{\partial v_m}{\partial x_j})$. Now this last expression can be computed using (3.3),

$$\begin{aligned}\delta_{-h}^i(\alpha g_{mj}\delta_h^i\frac{\partial v_m}{\partial x_j}) &= \delta_{-h}^i\delta_h^i(\alpha p) - \delta_{-h}^i(\delta_h^i(\alpha g_{mj})\tau_h^i\frac{\partial v_m}{\partial x_j})\\ &= \delta_{-h}^i\delta_h^i(\alpha p) - \delta_{-h}^i\delta_h^i(\alpha g_{mj})\frac{\partial v_m}{\partial x_j} - \delta_h^i(\alpha g_{mj})\delta_{-h}^i\tau_h^i\frac{\partial v_m}{\partial x_j}\\ &= \delta_{-h}^i\delta_h^i(\alpha p) - \delta_{-h}^i\delta_h^i(\alpha g_{mj})\frac{\partial v_m}{\partial x_j} + \delta_h^i(\alpha g_{mj})\delta_h^i\frac{\partial v_m}{\partial x_j}.\end{aligned}$$

Thus

$$(3.15)\quad \| \delta_{-h}^i(\alpha g_{mj}\delta_h^i\frac{\partial v_m}{\partial x_j}) \|_{L^2(G_{R'''})} \le c[\|\delta_h^i\alpha p\|_{H^1(G_{R''})} + \|v\|_{G^1(G_R)} + \| \delta_h^i v \|_{H^1(G_{R''})}]$$

Here we used the fact that αg_{mj} is C^2, Lemma 3.2 and the fact that the supports of v and ρ are actually in $G_{R'}$. Since $\delta^i_{-h}(\alpha g_{mj}\frac{\partial\varphi^{(\ell)}_m}{\partial x_j})$ converges weakly in $L^2(G_{R'''})$ as $\ell\to\infty$ to $\delta^i_{-h}(\alpha g_{mj}\delta^i_h\frac{\partial v_m}{\partial x_j})$ we may assume, by passing to a subsequence, if necessary, that their norms in $L^2(G_{R''})$ are uniformly bounded by the right-hand side of (3.15) (with a larger constant).

$$\|\delta^i_{-h}(\alpha g_{mj}\frac{\partial\varphi^{(\ell)}_m}{\partial x_j})\|_{L^2(G_{R'''})} \le c[\|\delta^i_h\alpha\rho\|_{H^1(G_{R''})} + \|v\|_{H^1(G_R)} + \|\delta^i_h v\|_{H^1(G_{R''})}]$$

Reading (3.14) for $\varphi^{(\ell)}$ and passing to $\limsup_{\ell\to\infty}$ we obtain

$$(3.16)\qquad |a(\delta^i_h v,\delta^i_h v)| \le c\{\|\delta^i_h v\|_{H^1(G_{R''})}[\|f\|_{L^2(G_R)} + \|v\|_{H^1(G_R)} + \|p\|_{L^2(G_R)}] + \|p\|_{L^2(G_R)}[\|v\|_{H^1(G_R)} + \|\delta^i_h(\alpha\rho)\|_{H^1(G_{R''})}]\}$$

Now, from the uniform ellipticity

$$a(\delta^i_h v,\delta^i_h v) \ge \frac{1}{M}\|\nabla\delta^i_h v\|^2_{L^2(G_{R''})} = \frac{1}{M}\|\delta^i_h v\|^2_{H^1(G_{R''})} - \frac{1}{M}\|\delta^i_h v\|^2_{L^2(G_{R''})}$$

and thus, with Lemma 3.2

$$(3.17)\qquad a(\delta^i_h v,\delta^i_h v) \ge \frac{1}{M}\|\delta^i_h v\|^2_{H^1(G_{R''})} - c\|v\|^2_{H^1(G_R)}\,.$$

Now we use (3.16), (3.17) and Young's inequality to deduce

$$(3.18)\qquad \|\delta^i_h v\|^2_{H^1(G_{R''})} \le c[\|v\|^2_{H^1(G_R)} + \|f\|^2_{L^2(G_R)} + \|p\|^2_{L^2(G_R)} + \|p\|_{L^2(G_R)}\|\delta^i_h(\alpha\rho)\|_{H^1(G_R)}]$$

The estimate (3.10) follows now from (3.18), Lemma 3.2 and Lemma 3.4. The proof of Lemma 3.6 is complete.

Let us consider now a bounded open set Ω with boundary $\partial\Omega$. Each point $y_0 \in \partial\Omega$ has a neighborhood U equipped with a C^3 diffeomorphism $\psi: U \to \mathbb{R}^n$ such that $U \cap \Omega$ is the preimage under ψ of the upper half plane $x_n > 0$ and $U \cap \partial\Omega$ is the preimage of $x_n = 0$. Restricting ψ and U to the preimage of a half ball in $\mathbb{R}^n$ containing $x_0 = \psi(y_0)$ it is easy to see that one can construct open sets U_j' such that $\overline{\Omega} \subset \bigcup_{j=1}^{r} U_j'$, $\overline{U_j'} \subset U_j$. There exists ψ_j defined in open neighborhoods of $\overline{U}_j$, C^3 diffeomorphism such that $\psi_j(U_j \cap \Omega) = G_{R_j}$, $\psi_j(U_j' \cap \Omega) = G_{R_j'}$, $0 < R_j' < R_j$. If $\overline{U_j'} \subset \Omega$ then G_{R_j} and $G_{R_j'}$ are balls. If $U_j' \cap \partial\Omega$ is not empty, then G_{R_j} and $G_{R_j'}$ are half balls. Let us consider a partition of unity $\alpha_j \in C_0^\infty(U_j')$, $0 \le \alpha_j \le 1$, $\sum_{j=1}^{r} \alpha_j = 1$, supp $\alpha_j \subset \overline{\Omega}$.

Let u,P be a weak solution of the Stokes system

$$-\nu\Delta u + \nabla P = F \tag{3.19}$$

$$\nabla\cdot u = 0 \tag{3.20}$$

in Ω. We asssume that $u \in H_0^1(\Omega)^n$ and $P \in L^2(\Omega)$, $F \in L^2(\Omega)^n$. We write $u = \sum_{j=1}^{r} \alpha_j u$, $P = \sum_{j=1}^{r} \alpha_j P$. Let us fix $j \in \{1,\dots,r\}$ and consider the pair $\alpha_j u$, $\alpha_j P$. We drop the index j for convenience.

Let us define

$$v(x) = (\alpha u)(\psi^{-1}(x)),\ p(x) = (\alpha P)(\psi^{-1}(x)). \tag{3.21}$$

Note first that the supports of v,q are included and compact in $G_{R'}$ ($G_{R'}$ in the case of half balls). Differentiating the identity $\alpha(y)u(y) = v(\psi(y))$ we get

$$\left(\alpha \frac{\partial u_i}{\partial y_k}\right)(y) = \frac{\partial \psi_m}{\partial y_k}(y)\,\frac{\partial v_i}{\partial x_m}(\psi^{-1}(y)) - \frac{\partial \alpha}{\partial y_k}(y)u_i(y) \tag{3.22}$$

and

(3.23) $$\left(\alpha \frac{\partial^2 u_i}{\partial y_k^2}\right)(y) = \frac{\partial \psi_{m'}}{\partial y_k} \frac{\partial \psi_m}{\partial y_k} \frac{\partial^2 v_i}{\partial x_m \partial x_{m'}} (\psi^{-1}(y)) + \frac{\partial^2 \psi_m}{\partial y_k^2}(y) \frac{\partial v_i}{\partial x_m}(\psi^{-1}(y)) - \frac{\partial^2 \alpha}{\partial y_k^2} u_i(y) - 2 \frac{\partial \alpha}{\partial y_k}(y) \frac{\partial u_i}{\partial y_k}(y).$$

We are lead to define thus

(3.24) $$g_{mk}(x) = \frac{\partial \psi_k}{\partial y_m} (\psi^{-1}(x))$$

(3.25) $$a_{ij}(x) = \nu \frac{\partial \psi_i}{\partial y_k} (\psi^{-1}(x)) \frac{\partial \psi_j}{\partial y_k} (\psi^{-1}(x))$$

Note that since ψ is a C^3 diffeomorphism, $g_{mk} \in C^2$ and (g_{mk}) is an invertible matrix. This implies property (3.4) for the matrix a_{ij}.

Taking the trace in (3.22) and using (3.20) we see that $\frac{\partial \psi_k}{\partial y_m} \frac{\partial v_m}{\partial x_k} = \frac{\partial \alpha}{\partial y_k} u_k$. Thus

(3.26) $$\rho(x) = g_{mk} \frac{\partial v_m}{\partial x_k} = \frac{\partial \alpha}{\partial y_i} (\psi^{-1}(x)) u_i(\psi^{-1}(x)).$$

We multiply (3.19) by α and use (3.23). We obtain an equation (3.8) for f appropriately defined

(3.27) $$\|f\|_{L^2(G_R)} \leq c[\|P\|_{L^2(\Omega)} + \|u\|_{H^1(\Omega)} + \|F\|_{L^2(\Omega)}]$$

We apply (3.18). We use for α in (3.18) the function $\tilde{\alpha}(x) = \alpha(\psi^{-1}(x))$. The reason we can do this is that $\rho(x) = \tilde{\alpha}(x) P(\psi^{-1}(x))$. Now from (3.26) we see that $\tilde{\alpha}(x)\rho(x) = \frac{\partial \alpha}{\partial y_i} (\psi^{-1}(x)) \cdot \tilde{\alpha}(x) u_i(\psi^{-1}(x))$. Now $\tilde{\alpha}(x) u_i(\psi^{-1}(x)) = v_i(x)$ by definition and thus

$$\tilde{\alpha}\rho = \beta \cdot v$$

where

$$\beta_i(x) = \frac{\partial \alpha}{\partial y_i} (\psi^{-1}(x)).$$

The need to consider $\tilde{\alpha}\rho$ instead of ρ arose from the fact that ρ itself is not an expression in v. Now

$$(3.28) \qquad \| \delta_h^i(\tilde{\alpha}\rho) \|_{H^1(G_{R''})} \le c[\| \delta_h^i v \|_{H^1(G_{R''})} + \|u\|_{H^1(\Omega)}].$$

Using (3.28) we can bootstrap in (3.18): We obtain

$$(3.29) \qquad \|\delta_h^i v\|^2_{H^1(G_{R''})} \le c[\|P\|^2_{L^2(\Omega)} + \|u\|^2_{H^1(\Omega)} + \|F\|^2_{L^2(\Omega)}]$$

where $v = (\alpha u)(\psi^{-1}(x))$ and $i = 1,2,\ldots,n-1$. Using Lemma (3.4) we infer

$$(3.30) \qquad \|D_i v\|^2_{H^1(G_R)} \le cE$$

where we denoted

$$(3.31) \qquad E = [\|P\|^2_{L^2(\Omega)} + \|u\|^2_{H^1(\Omega)} + \|F\|^2_{L^2(\Omega)}]$$

From (3.30) we know that D_iD_jv belongs to $L^2(G_R)$ for all i,j except $(i,j) = (n,n)$.

As in the case of a single elliptic equation we need to use the equation to get information about $\frac{\partial^2 v}{\partial x_n \partial x_n}$. In order to do so we first study the pressure. We shall use the expression "belongs to H^{-1}" (resp. "belongs to L^2") to mean that the corresponding quantity has a norm as a linear functional on H^1 (resp. norm in L^2) bounded by $c\sqrt{E}$ for an appropriate c.

We differentiate (3.8) in a tangential ($k < n$) direction. We infer that $g_{mj} \frac{\partial^2 p}{\partial x_k \partial x_j}$ belongs to $H^{-1}(G_R)$. (Indeed $\nabla(\frac{\partial^2 v}{\partial x_k \partial x_j})$ belongs to $H^{-1}(G_R)$ since $k < n$ and $\frac{\partial^2 v}{\partial x_k \partial x_j}$ belongs to L^2). Since g_{mj} is invertible, it follows that $\frac{\partial^2 p}{\partial x_\ell \partial x_k}$ belongs to $H^{-1}(G_R)$ for all $\ell = 1,2,\ldots,n$. But this implies

(3.32) $\frac{\partial p}{\partial x_k}$ belongs to $L^2(G_R)$, for all $k = 1,2,\ldots,n-1$.

Now let us take (3.26) and differentiate with respect to the normal direction $\frac{\partial}{\partial x_n}$. We infer that $g_{mn} \frac{\partial^2 v_m}{\partial x_n^2} + \sum_{k=1}^{n-1} g_{mk} \frac{\partial^2 v_m}{\partial x_n \partial x_k}$ belongs to $L^2(G_R)$. Using (3.30) we conclude that

(3.33) $$g_{mn} \frac{\partial^2 v_m}{\partial x_n^2} \text{ belongs to } L^2(G_R)$$

Let us return now to equation (3.8). Using (3.30) it follows that

(3.34) $$a_{nn} \frac{\partial^2 v_m}{\partial x_n^2} + g_{mj} \frac{\partial p}{\partial x_j} \text{ belongs to } L^2(G_R).$$

Using (3.32) it follows that

$$a_{nn} \frac{\partial^2 v_m}{\partial x_n^2} + g_{mn} \frac{\partial p}{\partial x_n} \text{ belongs to } L^2(G_R).$$

Multiplying by g_{mn} and summing in m we get

$$a_{nn} g_{mn} \frac{\partial^2 v_m}{\partial x_n^2} + (g_{mn} g_{mn}) \frac{\partial p}{\partial x_n} \text{ belongs to } L^2.$$

Using (3.33) we obtain that $(g_{mn} g_{mn}) \frac{\partial p}{\partial x_n}$ belongs to L^2.

Now since the matrix g_{mn} is invertible (actually $g_{mn} g_{mn} = a_{nn}$) it follows that

(3.35) $$\frac{\partial p}{\partial x_n} \text{ belongs to } L^2.$$

Now from (3.32), (3.34) and (3.35) it follows that $a_{nn} \frac{\partial^2 v_m}{\partial x_n^2}$ belongs to $L^2(G_R)$ and since $a_{nn} \neq 0$ finally

(3.36) $$\frac{\partial^2 v_m}{\partial x_n^2} \text{ belongs to } L^2(G_R).$$

In order to conclude, let us remark that (3.32) and (3.35) imply that $p \in H^1(G_R)$ and $\|p\|_{H^1(G_R)} \le c\sqrt{E}$. Similarly, (3.30) and (3.36) imply that $v \in H^2(G_R)$ and $\|v\|_{H^2(G_R)} \le c\sqrt{E}$. From the definitions of v and p it follows that $\alpha(y)u(y)$ and $\alpha(y)P(y)$ belong to $H^2(\Omega)$ and $H^1(\Omega)$ respectively (with norms bounded by $c\sqrt{E}$). Since $u = \sum \alpha_j u$, $P = \sum \alpha_j P$ we proved:

Theorem 3.7. Let Ω be a bounded open set with C^3 boundary. Let (u,p) solve the Stokes system weakly:

$$-\nu\Delta u + \nabla p = f \tag{3.37}$$

$$\nabla\cdot u = 0 \tag{3.38}$$

Assume $f \in L^2(\Omega)^n$, $u \in H_0^1(\Omega)^n$, $p \in L^2(\Omega)$. Then $u \in H^2(\Omega)^n$, $p \in H^1(\Omega)$. There exists a constant such that

$$\|u\|_{H^2(\Omega)} + \|p\|_{H^1(\Omega)} \le c[\|f\|_{L^2(\Omega)} + \|u\|_{H^1(\Omega)} + \|p\|_{L^2(\Omega)}] \tag{3.39}$$

Remark 3.8. If the boundary of Ω is assumed to be of class C^{2+m} and $f \in H^m(\Omega)$ then it can be shown that $u \in H^{m+2}(\Omega)^n$, $p \in H^{m+1}(\Omega)$ and

$$\|u\|_{m+2,\Omega} + \|p\|_{m+1,\Omega} \le c_m[\|f\|_{m,\Omega} + \|u\|_{1,\Omega} + \|p\|_{0,\Omega}] \tag{3.40}$$

The requirement that Ω be of class C^3 in Theorem 3.7 can be relaxed at the (academic) price of assuming $p \in H^1$, $u \in H^2$.

Theorem 3.9. Let Ω be a bounded open set of class C^2. Let (u,p) solve the Stokes system (3.37), (3.38) weakly. Assume $f \in L^2(\Omega)^n$, $u \in H_0^1(\Omega)^n \cap H^2(\Omega)^n$, $p \in H^1(\Omega)$. Then there exists a constant such that

$$\|u\|_{H^2(\Omega)} + \|p\|_{H^1(\Omega)} \le c[\|f\|_{L^2(\Omega)} + \|u\|_{H^1(\Omega)} + \|p\|_{L^2(\Omega)}] \tag{3.41}$$

Note the fact that the right-hand side of (3.41) does not contain $||p||_{H^1(\Omega)}$, $||u||_{H^2(\Omega)}$.

Theorem 3.9 cannot be used in conjunction with Theorem 2.3. One needs a constructive method or an approximation scheme that has $p \in H^1(\Omega)$, $u \in H^2(\Omega)$. These methods are available but will not be described here. Theorem 3.7, however, can be used together with Theorem 2.3 and provides existence and regularity of solutions to the Stokes system.

For the proof of Theorem 3.9 we need an analogue of Lemma 3.6.

Lemma 3.10. Let $0 < R' < R$. Let v,p be a weak solution of

$$(3.42) \qquad -\frac{\partial}{\partial x_i}\left(a_{ij}(x)\frac{\partial v_m}{\partial x_j}\right) + b_j(x)\frac{\partial v_m}{\partial x_j} + g_{mj}\frac{\partial p}{\partial x_j} = f_m, \quad m = 1,\dots,n$$

$$(3.43) \qquad g_{mk}\frac{\partial v_m}{\partial x_k} = \rho$$

where $a_{ij} \in C^1(G_R)$, $g_{mj} \in C^1(G_R)$, $b_j \in C^0(G_R)$, $f = (f_m) \in L^2(G_R)^n$, ρ $H^1(G_R)$. The principal part of (3.42) is assumed to be uniformly elliptic, (i.e., (3.4) holds). G_R is either a ball or a half ball. Assume that $v \in (H_0^1(G_{R'}))^n$, $p \in H^1(G_{R'})$ and that the supports of v,p are compact and contained in $G_{R'}$ ($\bar{G}_{R'}$ in the case of half balls). Then there exists a constant such that, for $i \le n$ (resp. $i \le n-1$ in the case of half balls)

$$(3.44) \qquad ||D^i v||^2_{H^1(G_R)} \le c[||f||^2_{L^2(G_R)} + ||p||^2_{L^2(G_R)} + ||v||^2_{H^1(G_R)}$$
$$+ ||D^i p||_{L^2(G_R)}||\rho||_{H^1(G_R)} + ||D^i p||_{L^2(G_R)}||v||_{H^1(G_R)}]$$

Proof. We proceed exactly as in the proof of Lemma 3.6. We estimate the term III = $\int \alpha p \frac{\partial}{\partial x_j}(g_{mj}\delta^i_{-h})dx$ differently. Using (3.3) one sees that III can be estimated by

$$|III| \le c[\|\delta^i_h(\rho)\|_{L^2(G_R)}\|\alpha g_{mj}\frac{\partial\varphi}{\partial x_j}\|_{L^2(G_{R''})} + \|p\|_{L^2(G_R)}\|\varphi\|_{H^1(G_R)}]$$

Now, if $\varphi^{(\ell)}$ converges to $\delta^i_h v$ in H^1 as $\ell \to \infty$, it follows that $\alpha g_{mj}\frac{\partial\varphi^{(\ell)}}{\partial x_j}$ converge in L^2 to $\alpha g_{mj}\delta^i_h\frac{\partial v_m}{\partial x_j} = \alpha\delta^i_h\rho - \alpha\delta^i_h(g_{mj})\tau^i_h(\frac{\partial v_m}{\partial x_j})$.
Therefore one obtains, passing to the limit, as $\ell \to \infty$

(3.45) $$|a(\delta^i_h v,\delta^i_h v)| \le c\{\|\delta^i_h v\|_{H^1(G_{R''})}[\|p\|_{L^2(G_R)} + \|f\|_{L^2(G_R)} + \|v\|_{H^1(G_R)}] + \|\delta^i_h p\|_{L^2(G_R)}\|\alpha\delta^i_h\rho\|_{L^2(G_R)} + \|\delta^i_h p\|_{L^2(G_R)}\|v\|_{H^1(G_R)}\}$$

Using the coercivity of a(. , .) we get

(3.46) $$\|\delta^i_h v\|^2_{H^1(G_{R''})} \le c[\|p\|^2_{L^2(G_R)} + \|f\|^2_{L^2(G_R)} + \|v\|^2_{H^1(G_R)} + \|\delta^i_h p\|_{L^2(G_R)}\|v\|_{H^1(G_R)} + \|\delta^i_h p\|_{L^2(G_R)}\|\delta^i_h(\alpha\rho)\|_{L^2(G_R)}]$$

The estimate (3.44) follows from (3.46) by Lemma 3.2.

In order to proceed let us assume that $g_{kj}(x) = \delta_{kj}$ if $j < n$, $g_{nn}(x) = 1$ and that $g_{kn} \in C^1$, $\frac{\partial g_{kn}}{\partial x_n} = 0$ (C^1 would suffice). Thus, from (3.42) we get

(3.47) $$\frac{\partial p}{\partial x_m} = f_m + \frac{\partial}{\partial x_i}(a_{ij}\frac{\partial v_m}{\partial x_j}) + b_j\frac{\partial v_m}{\partial x_j} - g_{mn}\frac{\partial p}{\partial x_n}, \quad m < n$$

(3.48) $$\frac{\partial p}{\partial x_n} = f_n + \frac{\partial}{\partial x_i}(a_{ij}\frac{\partial v_n}{\partial x_j}) + b_j\frac{\partial v_n}{\partial x_j}.$$

Differentiating (3.48) with respect to some tangential direction $\frac{\partial}{\partial x_k}$, $k < n$ (if G_R is a ball $k \leq n$) then we see that

$$\| \frac{\partial^2 p}{\partial x_k \partial x_n} \|_{H^{-1}(G_R)} \leq c[\| f\|_{L^2(G_R)} + \| \frac{\partial v}{\partial x_k} \|_{H^1(G_R)} + \|v\|_{H^1(G_R)}]$$

(Only first order derivatives of a_{ij} and no derivatives on b are used.)

We differentiate (3.47) with respect to x_k. The only interesting term is

$$\frac{\partial}{\partial x_k} (g_{mn} \frac{\partial p}{\partial x_n}) = \frac{\partial}{\partial x_n} (g_{mn} \frac{\partial p}{\partial x_k} + \frac{\partial g_{mn}}{\partial x_k} p) = g_{mn} \frac{\partial^2 p}{\partial x_n \partial x_k} + \frac{\partial}{\partial x_n} (\frac{\partial g_{mn}}{\partial x_k} p)$$

It follows that

$$\|\frac{\partial^2 p}{\partial x_m \partial x_k} \|_{H^{-1}(G_R)} \leq c[\|f\|_{L^2(G_R)} + \| \frac{\partial v}{\partial x_k} \|_{H^1(G_R)} + \|v\|_{H^1(G_R)} + \|p\|_{L^2(G_R)}]$$

Using the estimate $\| \frac{\partial p}{\partial x_k} \|_{L^2(G_{R''})} \leq c\| \nabla \frac{\partial p}{\partial x_k} \|_{H^{-1}(G_R)}$ valid since the mean of $\frac{\partial p}{\partial x_k}$, $\int_{G_R} \frac{\partial p}{\partial x_k}$ equals 0, we obtain

$$(3.49) \quad \|D_k p\|_{L^2(G_R)} \leq c[\| f\|_{L^2(G_R)} + \|D_k v\|_{H^1(G_R)} + \|v\|_{H^1(G_R)} + \| p\|_{L^2(G_R)}]$$

Combining (3.44) and (3.49) we obtain, after a bootstrap

$$(3.50) \quad \|D_k v\|^2_{H^1(G_R)} \leq c[\| f\|^2_{L^2(G_R)} + \| p\|^2_{L^2(G_R)} + \|v\|^2_{H^1(G_R)} + \|\rho\|^2_{H^1(G_R)}]$$

Now the proof of Theorem 3.9 will follow from (3.50) and the proof of Theorem 3.7 provided we are able to produce g_{mj} with the special form. If Ω is of class C^2 then, locally, after a rotation and a translation if necessary, Ω is given by $y_n > \sigma(y')$ for some $\sigma \in C^2$, $y' = (y_1,\ldots,y_{n-1})$. Therefore after localization and a rigid motion (translation + rotation) the change of variables which achieves the flattening of the boundary $\psi(y)$ is given by $\psi(y) = (y', y_n - \sigma(y'))$. Now, the Stokes system is invariant to rigid motions. The functions $g_{kj}(x) = \frac{\partial \psi_j}{\partial y_k}(x', x_n + \sigma(x'))$ satisfy the required properties $g_{kj} = \delta_{kj}$ for $j < n$, $g_{nn} = 1$, $g_{jn} = -\frac{\partial \sigma}{\partial x_j}(x')$ C^1, $j < n$, $\frac{\partial g_{jn}}{\partial x_n} = 0$.

Theorem 3.11. Let Ω be open bounded of class C^2. Let $f \in L^2(\Omega)^n$. There exists and are unique $u \in H^2(\Omega) \cap V$, $p \in H^1(\Omega)$ solutions of the Stokes system

$$\begin{cases} -\nu\Delta u + \operatorname{grad} p = f & \text{in } \Omega \\ \operatorname{div} u = 0 & \text{in } \Omega \\ u\big|_{\partial\Omega} = 0 \end{cases}$$

Moreover,

$$\|u\|_{H^2(\Omega)} + \|p\|_{H^1(\Omega)/R} \leq c\|f\|_{L^2(\Omega)} \tag{3.51}$$

The proof, in the case Ω of class C^3, consists of Theorem 2.3 and Theorem 3.7. In the case Ω of class C^2 one uses Theorem 3.9 (or rather its proof).

4
THE STOKES OPERATOR

We recall that we denoted by P the orthogonal projection $P: L^2(\Omega)^n \to H$. Let us assume that Ω is bounded, $\partial\Omega$ of class C^2.

Definition 4.1. The Stokes operator A is defined by

(4.1) $$A: \mathcal{D}(A) \subset H \to H, \quad A = -P\Delta, \quad \mathcal{D}(A) = H^2(\Omega) \cap V.$$

Proposition 4.2. The Stokes operator is symmetric, i.e.,

(4.2) $$(Au,v) = (u,Av) \quad \text{for all } u,v \in \mathcal{D}(A).$$

Proof. Let us assume first that $u,v \in (C_0^\infty(\Omega))^n$ and $\operatorname{div} u = \operatorname{div} v = 0$. Then, since $Pu = u$, $Pv = v$, (4.2) is nothing but the familiar

(4.3) $$-\int_\Omega (\Delta u_i) v_i \, dx = \int_\Omega \frac{\partial u_i}{\partial x_j} \frac{\partial v_i}{\partial x_j} \, dx .$$

Now, if u,v are in $\mathcal{D}(A)$ and arbitrary we can approximate them in $H^1(\Omega)^n$ by functions in V. If $u \in \mathcal{D}(A)$ and $v \in V$ then obviously (4.3) holds. Passing to the limit in the v's in $H^1(\Omega)$ we get (4.3) for arbitrary $u \in \mathcal{D}(A)$, $v \in V$. In particular (4.3) means

(4.4) $$(Au,v) = ((u,v)) \quad \text{for all } u,v \in \mathcal{D}(A)$$

Since the right hand side of (4.4) is symmetric the Proposition is proven. We note that (4.4) is true for $u \in \mathcal{D}(A)$, $v \in V$.

Theorem 4.3. The Stokes operator is selfadjoint.

Proof. Let u be an element of $\mathcal{D}(A^*)$. By definition there exists $f \in H$ such that

$$(Av,u) = (v,f) \quad \text{for all } v \in \mathcal{D}(A)$$

Since $f \in H \subset L^2(\Omega)^n$ we can find, by Theorem 3.12, $\tilde{u}, p$, $\tilde{u} \in \mathcal{D}(A)$ such that $A\tilde{u} = f$. We want to show that $u = \tilde{u}$. In order to do so let us compute $(g, u - \tilde{u})$ for arbitrary $g \in H$. Using Theorem 3.12 again, there exists $v \in \mathcal{D}(A)$ such that $Av = g$. Therefore

$$(g,u-\tilde{u}) = (Av,u) - (Av,\tilde{u}) = (v,f) - (v,A\tilde{u}) = (v,f) - (v,f) = 0.$$

Since $g \in H$ is arbitrary $u = \tilde{u}$ and thus $u \in \mathcal{D}(A)$ and $f = Au$.

Theorem 4.4. The inverse of the Stokes operator, A^{-1}, is a compact operator in H.

Proof. For $f \in H$, $A^{-1}f = u$, where u is the unique solution in $H^2(\Omega) \cap V = \mathcal{D}(A)$ of the Stokes equation. In view of (3.51) $A^{-1}: H \to V$ is bounded. Since the inclusion $V \subset H$ is compact by Rellich's Theorem, the result follows.

Now $K = A^{-1}$ is self-adjoint, injective and compact. By a well known theorem of Hilbert, there exists a sequence of positive numbers $\mu_j > 0$, $\mu_{j+1} \leq \mu_j$ and an orthonormal basis of H, (w_j) such that $Kw_j = \mu_j w_j$. We denote $\lambda_j = \mu_j^{-1}$. Since A^{-1} has range in $\mathcal{D}(A)$ we obtain that

(4.4) $\quad Aw_j = \lambda_j w_j, \quad w_j \in \mathcal{D}(A)$

(4.5) $\quad 0 < \lambda_1 < \cdots \leq \lambda_j \leq \lambda_{j+1} \leq \cdots$

(4.6) $$\lim_{j\to\infty} \lambda_j = \infty$$

(4.7) $(w_j)_{j=1,\dots}$ are an orthonormal basis of H.

Proposition 4.5. If Ω is bounded of class $C^{\ell+2}$, $\ell \geq 0$ then $w_j \in H^{\ell+2}(\Omega)^n$.

The proof follows from Remark 3.8.

Let $\alpha > 0$ be a real number. We define the operator A^α by

(4.8) $$A^\alpha u = \sum_{j=1}^{\infty} \lambda_j^\alpha u_j w_j \quad \text{for} \quad u = \sum_{j=1}^{\infty} u_j w_j, \quad u \in \mathcal{D}(A^\alpha)$$

(4.9) $$\mathcal{D}(A^\alpha) = \{u \in H \mid u = \sum_{j=1}^{\infty} u_j w_j, \ \sum_{j=1}^{\infty} \lambda_j^{2\alpha} |u_j|^2 < \infty, \ u_j \in \mathbb{R}\}$$

The spaces $\mathcal{D}(A^\alpha)$ carry a natural scalar product

(4.10) $$\langle u,v\rangle_\alpha = \sum_{j=1}^{\infty} \lambda_j^{2\alpha} u_j v_j \quad \text{if} \quad u = \sum_{j=1}^{\infty} u_j w_j, \quad v = \sum_{j=1}^{\infty} v_j w_j$$

For this scalar product, the vectors $\lambda_j^{-\alpha} w_j$, $j = 1,\dots$ form an orthonormal system which is complete. For $\alpha = 1/2$, we have $\mathcal{D}(A^{1/2}) = V$ and $\langle u,v\rangle_1 = ((u,v))$. Indeed the vectors $\lambda_j^{-1/2} w_j$ belong to V and on them

$$\langle \lambda_j^{-1/2} w_j, \lambda_k^{-1/2} w_k\rangle_{1/2} = \delta_{jk} \lambda_j^{1} \cdot \lambda_j^{-1} = \delta_{jk} = (A(\lambda_j^{-1/2} w_j), \lambda_k^{-1/2} w_j)$$
$$= ((\lambda_j^{-1/2} w_j, \lambda_k^{-1/2} w_k)).$$

Therefore $\mathcal{D}(A^{1/2}) \subset V$ and it is a closed subspace of V. If $v \in V$ is orthogonal in V to $\mathcal{D}(A^{1/2})$ then, in particular, $((v,w_j)) = 0$ for all j. But $v \in V$, $w_j \in \mathcal{D}(A)$ and from (4.4) (and the observations following it)

$$0 = ((v,w_j)) = (v,Aw_j) = \lambda_j(v,w_j).$$

Since $\lambda_j > 0$ it follows that $(v,w_j) = 0$ for all j and thus $v = 0$.

Proposition 4.6. $\mathcal{V} \subset \mathcal{D}(A^\alpha)$ for all $\alpha > 0$. In particular, $\bigcap_{\alpha > 0} \mathcal{D}(A^\alpha)$ is dense in H.

Proof. Clearly $\mathcal{D}(A^\alpha) \subset \mathcal{D}(A^\beta)$ because $(\lambda_j/\lambda_1)^{\alpha-\beta} > 1$ if $\alpha > \beta$, $j = 1,2,\dots$. It is enough therefore to show $\mathcal{V} \subset \mathcal{D}(A^p)$ for any positive integer p. Let $\varphi \in \mathcal{V}$. Then $(-\Delta)\varphi \in \mathcal{V}$. In particular, it is divergence free and thus $P(-\Delta)\varphi = -\Delta\varphi$. So $A\varphi = (-\Delta)\varphi$ for $\varphi \in \mathcal{V}$. (Incidentally, we showed that $Au = -\Delta u$ for $u \in H_0^2(\Omega)^n \cap V$ or in other words that $Au \neq -\Delta u$ may happen only for those $u \in \mathcal{D}(A)$ which do not belong to $H_0^2(\Omega)^n$.) Therefore for $\varphi \in \mathcal{V}$, $A^p\varphi \in \mathcal{V}$. Now

$$\lambda_j^p(w_j, \varphi) = (A^p w_j, \varphi) = (w_j, A^p\varphi)$$

and since $A^p\varphi \in H$, the series of its Fourier coefficients is square summable:

$$\sum_{j=1}^{\infty} \lambda_j^{2p} |(w_j, \varphi)|^p = \sum_{j=1}^{\infty} |(w_j, A^p\varphi)|^2 < \infty$$

This means $\varphi \in \mathcal{D}(A^p)$. Moreover, we proved

$$\langle u,u\rangle_p \leq |(-\Delta)^p u|^2_{L^2(\Omega)^n} \quad \text{for all } u \in \mathcal{V},\ p \in \mathbb{N}. \tag{4.11}$$

Remark though that $\mathcal{V}$ may not be dense in $\mathcal{D}(A^p)$ with the $\langle\ ,\ \rangle_p$ topology, as one can see in the case $p = 1$.

Before addressing the question of the relationship between spaces $\mathcal{D}(A^m)$ and $H^{2m}(\Omega)$ we discuss the scale dependence of various quantities.

We shall say that a function of the set Ω, $c(\Omega)$, is scale invariant if $c(\Omega) = c(\Omega')$ for all Ω' obtained from Ω by a rigid motion and a dilation $x \to \delta x$. Let us denote by T_δ and Ω_δ the operation transporting functions defined on Ω, f(x), to functions defined on $\Omega_\delta = \{\delta x | x \in \Omega\}$, $(T_\delta f)(y) = f(\frac{y}{\delta})$. When we refer to the way something scales we mean under the dilations δ and operations T_δ. One can easily modify the definition of the norms $\| \ \|_{m,\Omega}$ in such a way that they scale as the pure m-th order derivatives do. We shall denote by $|\Omega| = \int_\Omega 1\, dx$ and by $L(\Omega) = L = |\Omega|^{1/n}$ the linear size of Ω. Then the natural definition of $\| \ \|_{m,\Omega}$ is

$$\|f\|^2_{m,\Omega} = \sum_{|\alpha| \le m} |\Omega|^{\frac{2(|\alpha|-m)}{n}} \int_\Omega |D^\alpha f|^2 dx \tag{4.13}$$

With this definition the quantity $\|f\|_{m,\Omega}$ scales like $L^{\frac{n}{2}-m}$, i.e.,

$$\frac{\|f\|_{m,\Omega}}{|\Omega|^{\frac{1}{2}-\frac{m}{n}}} \quad \text{is scale invariant.} \tag{4.14}$$

Clearly the operator A scales like L^{-2} and thus

$$|\Omega|^{\frac{2}{n}} \lambda_j \quad \text{is scale invariant, } j = 1,\ldots \tag{4.15}$$

and of course

$$\frac{\lambda_j}{\lambda_1} \quad \text{is scale invariant, } j = 1,\ldots \tag{4.16}$$

Finally if $w_j(x)$ is an orthonormal basis formed with eigenfunctions of A in the domain Ω, then in Ω_δ the corresponding orthonormal basis will be

$$w_j^\delta(y) = \delta^{-\frac{n}{2}} w_j(\tfrac{y}{\delta}).$$

So w_j scales like $L^{-n/2}$:

$$L^{n/2} w_j = |\Omega|^{1/2} w_j \quad \text{is scale invariant.} \tag{4.17}$$

Proposition 4.7. Let Ω be bounded open of class C^2. Then there exists a constant (scale invariant) such that

$$(4.18) \qquad \|u\|_{2,\Omega} \le C(\Omega)|Au|, \quad \text{for all } u \in \mathcal{D}(A).$$

Proof. If $u \in \mathcal{D}(A)$ then $Au = f \in H \subset L^2(\Omega)^n$ and $u \in H^2(\Omega)^n$. Therefore (4.18) is a consequence of Theorems 2.3 and 3.9.

We are going to investigate now the size of the eigenvalues λ_j.

Lemma 4.8. Let $\Omega \subset \mathbb{R}^n$ be open, bounded and of class C^ℓ. There exists $E^\ell_\Omega : H^m(\Omega) \to H^m(\mathbb{R}^n)$ bounded linear operator, $E^\ell_\Omega(u)\big|_\Omega = u$, for all $u \in H^m(\Omega)$, $m \le \ell$.

Proof. Consider first the case of $R^n_+ = \{x \in \mathbb{R}^n \mid x_n > 0\}$. Let us define for $x_n \ge 0$ and $u \in C^\ell(\overline{\Omega})$

$$E^\ell(u)(x',-x_n) = \sum_{j=1}^{\ell+1} a_j u(x',jx_n)$$

where a_j are defined by

$$(-1)^k = \sum_{j=1}^{\ell+1} j^k a_j \qquad k = 0,\dots,\ell$$

Such numbers obviously exist because $(j^k)_{\substack{j=1,\dots,\ell+1\\ k=0,1,\dots,\ell}}$ has a Vandermonde determinant $= \prod_{1\le i<j\le \ell+1} (j-i) \ne 0$.

Clearly the function $E^\ell(u)$ is continuous, together with ℓ derivatives across $x_n = 0$. Moreover

$$\|E^\ell(u)\|_{H^m(\mathbb{R}^n)} \le c_m \|u\|_{H^m(R^n_+)} \quad \text{for all } m \le \ell$$

obviously. Since the functions belonging to $C^\ell(R^n_+)$ and having finite $H^m(R^n_+)$ are dense in $H^m(R^n_+)$ we can extend $E^{(\ell)}$ by continuity. The case of a general domain is reduced to that of a halfspace by a partition of unity.

Lemma 4.9 Let $0 < s_1 < n/2 < s_2$. Define $t \in (0,1)$ such that $n/2 = (1 - t)s_1 + ts_2$ (i.e., $t = (\frac{n}{2} - s_1)/(s_2 - s_1)$). Then there exists a constant $c = c(n,s_1,s_2)$ depending on n,s_1,s_2 only such that, for every $f \in H^{s_2}(\mathbb{R}^n)$

$$\|f\|_{L^\infty} \le c\|f\|_{s_1}^{1-t} \|f\|_{s_2}^{t} \tag{4.19}$$

Proof. Let $f \in C_0^\infty$. Then

$$f(x) = (2\pi)^{-n}\int e^{i<x,\xi>} \hat{f}(\xi)d\xi = (2\pi)^{-n}\int_{|\xi|<R} + (2\pi)^{-n}\int_{|\xi|>R}$$

$$\int_{|\xi|<R} |\hat{f}(\xi)|d\xi = \int_{|\xi|<R} (1+|\xi|^2)^{-s_1/2}(1+|\xi|^2)^{s_1/2}|\hat{f}(\xi)|d\xi$$
$$\le c_{n,s_1} R^{\frac{n-2s_1}{n}} \|f\|_{s_1}$$

with $c_{n,s_1} = \sqrt{\frac{\omega_n}{n-2s_1}}$, ω_n the area of S^{n-1}.

$$\int_{|\xi|>R} |\hat{f}(\xi)|d\xi = \int_{|\xi|>R} (1+|\xi|^2)^{-s_2/2}(1+|\xi|^2)^{s_2/2}|\hat{f}(\xi)|d\xi$$
$$\le c_{n,s_2} R^{\frac{n-2s_2}{2}} \|f\|_{s_2}$$

with $c_{n,s_2} = \sqrt{\frac{\omega_n}{2s_2 - n}}$. Equating $R^{\frac{n}{2} - s_1} \|f\|_{s_1} = R^{\frac{n}{2} - s_2} \|f\|_{s_2}$ gives the choice $R = (\frac{\|f\|_{s_2}}{\|f\|_{s_1}})^{1/(s_2-s_1)}$ and yields (4.19) with

$$C = \sqrt{\omega_n}(\frac{1}{\sqrt{2s_2- n}} + \frac{1}{\sqrt{n - 2s_1}}).$$

Lemma 4.10. Let Ω be an open, bounded set of class C^ℓ. Assume $\ell > n/2$. Let ℓ' be an integer $\ell' < n/2$. Then there exists a constant depending on Ω (scale invariant), ℓ, ℓ' such that, for any $f \in H^\ell(\Omega)$

(4.20) $$\|f\|_{L^\infty(\Omega)} \le C\|f\|_{\ell',\Omega}^{1-t}\|f\|_{\ell,\Omega}^{t}$$

where $t = \dfrac{\frac{n}{2} - \ell'}{\ell - \ell'}$.

Proof. Let $E_\Omega^\ell(f)$ be the extension of f to $H^\ell(\mathbb{R}^n)$

$$\| E_\Omega^\ell f \|_\ell \le c_\ell \|f\|_{\ell,\Omega}, \quad \| E_\Omega^\ell f \|_{\ell'} \le c_{\ell'}\|f\|_{\ell',\Omega}.$$

$\|f\|_{L^\infty(\Omega)} \le \| E_\Omega^\ell f \|_{L^\infty(R^n)}$ since $E_\Omega^\ell f\big|_\Omega = f$. This proves (4.20) as a consequence of (4.19). For the scale invariance we could proceed in two ways. We could check the scale independence of each step in the proof (c_ℓ, $c_{\ell'}$ in particular). This is clear, if one pauses a moment to think about the proof of Lemma 4.8. Or, we can check that the expression $\|f\|_{L^\infty(\Omega)}/\|f\|_{\ell',\Omega}^{1-t}\|f\|_{\ell,\Omega}^{t}$ is scale invariant. Now $\|f\|_{L^\infty(\Omega)}$ is scale invariant and $\|f\|_{\ell',\Omega}^{1-t}\|f\|_{\ell,\Omega}^{t}$ scales like

$$L^{(1-t)(\frac{n}{2} - \ell')} L^{t(\frac{n}{2} - \ell)} = L^{\frac{n}{2} - \frac{n}{2}} = L^0.$$

Theorem 4.11. Let $\Omega \subset \mathbb{R}^n$, n = 2 or 3 be open bounded and of class C^2. There exists a scale invariant constant c_0 such that, for every $j = 1,2,\ldots$ the eigenvalues λ_j of the Stokes operator satisfy

(4.21) $$\lambda_j \ge c_0 j^{2/n}\lambda_1.$$

Proof. Let $w_k(x)$ be the sequence of the corresponding eigenfunctions. Let $\alpha_1,\ldots,\alpha_j$ be arbitrary real numbers and let $w = \sum_{k=1}^{j} \alpha_k w_k$. Thus, applying (4.20) with $\ell' = 0$ and consequently $t = \frac{n/2}{2} = \frac{n}{4}$ we get

$$\|w\|_{L^\infty(\Omega)} \le c_1 |w|^{\frac{4-n}{4}} \|w\|_{2,\Omega}^{n/4}$$

By Proposition (4.18), $\|w\|_{2,\Omega} \le c_2 |Aw|$ so $\|w\|_{L^\infty} \le c_3 |w|^{\frac{4-n}{4}} |Aw|^{n/4}$. But

$|w|^2 = \sum_{k=1}^{j} \alpha_k^2$ because w_k are orthonormal. Also

$|Aw|^2 = \sum_{k=1}^{j} \lambda_k^2 \alpha_k^2 \le \lambda_j^2 (\sum_{k=1}^{j} \alpha_k^2)$ because w_k are eigenfunctions and

$\lambda_k \le \lambda_{k+1}$. It follows that $\|w\|_{L^\infty(\Omega)} \le c_3 \lambda_j^{n/4} (\sum_1^j \alpha_k^2)$. It follows that $|w(x)|^2 \le c_4 \lambda_j^{n/2} (\sum_1^j \alpha_k^2)$ almost everywhere. Actually, since $H^2(\Omega) \subset C^0(\Omega)$, the inequality holds for all x. Let $1 \le i \le n$. Let us denote by $w_k^{(i)}(x)$ the i-th component of the vector $w_k(x)$. Then

$$\left|\sum_{k=1}^{j} \alpha_k w_k^{(i)}(x)\right|^2 \le |w(x)|^2 \le c_4 \lambda_j^{n/2} (\sum_1^j \alpha_k^2).$$

Choosing $\alpha_k = w_k^{(i)}(x)$ we obtain

$$\sum_{k=1}^{j} |w_k^{(i)}(x)|^2 \le c_4 \lambda_j^{n/2}.$$

Summing in i

$$\sum_{k=1}^{j} |w_k(x)|^2 \le n c_4 \lambda_j^{n/2}, \quad \text{for all } x \in \Omega.$$

Integrating over Ω we obtain $j \le n c_4 \lambda_j^{n/2} |\Omega|$ and thus

$$\lambda_j \ge c|\Omega|^{-2/n}\, j^{2/n}. \tag{4.22}$$

The scale invariance follows from the scale invariance of the constants $c_1, c_2, \ldots$ and from (4.15), (4.16).

<u>Proposition 4.12</u>. Let $\Omega \subset \mathbb{R}^n$, n = 2 or 3 be an open bounded set of class C^ℓ, $\ell \ge 2$. Then

$$H^m(\Omega)^n \cap H_0^{m-1}(\Omega)^n \cap V \subset \mathcal{D}(A^{m/2}), \quad \text{if } m \text{ is even} \tag{4.23}$$

(4.24) $$H_0^m(\Omega)^n \cap V \subset \mathcal{D}(A^{m/2}), \quad \text{if } m \text{ is odd}$$

(4.25) $$\mathcal{D}(A^{m/2}) \subset (H^m(\Omega))^n \cap V \quad \text{provided } m \geq 1,\ m \leq \ell.$$

Proof. Let $u \in H^m(\Omega)^n \cap H_0^{m-1}(\Omega)^n \cap V$, $m = 2p$. Then $(-\Delta)^k u \in H_0^{m-1-2k}(\Omega)^n$ for $k = 0,\ldots,p-1$. Moreover, $\operatorname{div}(-\Delta)^k u = 0$. Therefore $(-\Delta)^k u \in H$, $P(-\Delta)^k u = (-\Delta)^k u$, $k \leq p-1$. Let w_j be an eigenfunction. Then

$$\lambda_j^{m/2}(u,w_j) = (u,A^p w_j) = (u,P(-\Delta)P(-\Delta)\cdots P(-\Delta)w_j) = ((-\Delta)^{p-1}u,-\Delta w_j).$$

Now since $(-\Delta)^{p-1}u \in H^2(\Omega)^n \cap H_0^2(\Omega)^1$ and since $w_j \in H^2(\Omega)^n \cap H_0^1(\Omega)^n$, integrating twice by parts we get

$$\lambda_j^{m/2}(u,w_j) = ((-\Delta)^p u,w_j)$$

and thus

$$\sum \lambda_j^m |(u,w_j)|^2 \leq \| (-\Delta)^p u \|_{0,\Omega}^2 \leq \|u\|_{m,\Omega}^2$$

or

(4.26) $$\langle u,u\rangle_{m/2} \leq \|u\|_{m,\Omega}^2$$

for all $u \in H_0^{m-1}(\Omega)^n \cap H^m(\Omega) \cap V$, if $m = 2p$. If $m = 2p+1$ for some integer $p \geq 0$ then we assume that $u \in H_0^m(\Omega)^n$, $\operatorname{div} u = 0$. Then $(-\Delta)^k u \in H_0^{m-2k}(\Omega)^n$, $k = 0,1,\ldots,p$ and thus $P(-\Delta)^k u = (-\Delta)^k u$, $k = 0,1,\ldots,p$. Then

$$\lambda_j^{m/2}(u,w_j) = (u,A^p A^{1/2} w_j) = ((-\Delta)^p u,A^{1/2}w_j) = (A^{1/2}(-\Delta)^p u,w_j).$$

Now $(-\Delta)^p u \in H_0^1(\Omega)^n$, $\operatorname{div}(-\Delta)^p u = 0$ and therefore $(-\Delta)^p u \in V$ and thus $|A^{1/2}((-\Delta)^p u)| = \| (-\Delta)^p u \| \leq \|u\|_{m,\Omega}$. The rest follows as in the even case: (4.26) is true for $m = 2p+1$, $u \in H_0^m(\Omega)^n \cap V$.

Let us prove now (4.25). Let $u \in \mathcal{D}(A^{m/2})$. We know, for $m = 0$ and $m = 1$ that

$$(4.27) \qquad \|u\|_{m,\Omega} \le c(\Omega)|A^{m/2}u|, \quad \text{for all } u \in \mathcal{D}(A^{m/2}).$$

Suppose by induction that (4.27) is true for $m' \le m-1$, $m' \ge 1$. Let $u \in \mathcal{D}(A^{m/2})$. Then $Au = f \in \mathcal{D}(A^{(m-2)/2})$. Moreover $u \in V$. Since $\|f\|_{m-2} \le c(\Omega)|A^{m/2}u|$ by (4.27) for m-2 it follows from (3.40) that (4.27) is true. The proof is complete.

The difference between the Laplacian for the Dirichlet problem and the Stokes operator originates from the fact that Leray's projector P and $(-\Delta)$ do not commute, in general. In the absence of boundaries, however, P and $-\Delta$ commute. By absence of boundaries we mean either the case $\Omega = \mathbb{R}^n$ and usual Sobolev spaces or the case $\Omega = \mathbb{T}^n$ the n-th dimensional torus. Let us describe the latter situation. Let $L > 0$ be a real. We denote by $Q_L = \{x \in \mathbb{R}^n \mid |x_i| < L/2\} = (-L/2, L/2)^n$. For each $k \in \mathbb{Z}^n$, $k \ne 0$ we define

$$(4.28) \qquad w_{k,j}(x) = L^{-n/2}\, e^{\frac{2\pi i}{L}\langle k,x\rangle} e_j$$

where $e_j = (\delta_{kj})_{k=1,\dots,n}$ are the canonical basis in $\mathbb{R}^n$, $i = \sqrt{-1}$, and we define, for $\alpha, \beta \in \mathbb{C}^n$,

$$(4.29) \qquad \langle \alpha, \beta\rangle = \sum_{j=1}^{n} \alpha_j \beta_j \ .$$

(Note the absence of complex conjugate in 4.29.) We shall consider complex valued periodic functions of period L. We shall denote by $w_k(x)$ the vector $w_k(x) = (w_{k,j})_{j=1,\dots,n}$. Then a vector valued periodic function u will have an expansion

$$(4.30) \qquad u \sim \sum_{k \in \mathbb{Z}^n} u_k w_k \ , \quad \text{where } u_k \in \mathbb{C}^n.$$

If we want u to be real we need to impose $\overline{u_k} = u_{-k}$ for all $k \in Z^n$. For each $\alpha > 0$ we define, for a periodic function u

(4.31) $$\|u\|_{\alpha,L} = L^{-\alpha}\Big(\sum_{k \in Z^n} (1 + |k|^2)^{\alpha}|u_k|^2\Big)^{1/2}$$

where

(4.32) $$u_k = \int_{Q_L} \langle u, w_{-k}\rangle dx.$$

Note that u_k scales like $L^{d/2}$ so (4.31) agrees with (4.14). The space of functions u such that $\|u\|_{\alpha,L} < \infty$ is denoted $H_{\alpha,L}$.

We define H,V now by the conditions

(4.33) $$H = \{u \in H_{0,L} \mid \overline{u}_k = u_{-k},\ u_0 = 0,\ \langle u_k, k\rangle = 0\}$$

(4.34) $$V = \{u \in H_{1,L} \mid \overline{u}_k = u_{-k},\ u_0 = 0,\ \langle u_k, k\rangle = 0\}$$

We set $A: \mathcal{D}(A) \to H$.

(4.35) $$\mathcal{D}(A) = H_{2,L} \cap V = H_{2,L} \cap H$$

(4.36) $$Au = \sum_{k \in Z^n} \frac{4\pi^2}{L^2} |k|^2 u_k w_k$$

i.e.,

(4.37) $$(Au)_k = \frac{4\pi^2}{L^2} |k|^2 u_k.$$

We note that Leray's projector P acts on components

(4.38) $$(Pu)_k = P_k u_k$$

(4.39) $$P_k u_k = u_k - \langle u_k, \frac{k}{|k|}\rangle \frac{k}{|k|} \quad \text{for } k \in Z^n \setminus 0$$

(4.40) $$P_0 u_0 = 0 \quad \text{at } k = 0.$$

Clearly the operator $-\Delta$ is given by (4.36) only that its domain is not restricted to div $u = 0$ elements.

Remark 4.13. The projector P commutes with Δ.

In this case, the spaces $\mathcal{D}(A^{\alpha/2})$ are easily identifiable

(4.41) $$\mathcal{D}(A^{\alpha/2}) = H_{\alpha,L} \cap V = \{u \in H_{\alpha,L} \mid \overline{u}_k = u_{-k},\ u_0 = 0,\ \langle u_k, k\rangle = 0\}.$$

The eigenvalues of A are

(4.42) $$\{\lambda_m\}_{m=1,2,\ldots} = \{\frac{4\pi^2}{L^2} |k|^2\}_{k \in Z^n \setminus \{0\}}$$

Clearly, for each $k \in Z^n \setminus \{0\}$ there are $2(n-1)$ eigenfunctions corresponding to it: they are of the form $cw_k + \overline{c}w_{-k}$ where the vector $c \in \mathbb{C}^n$ satisfies the equations

$$\langle c, k\rangle = 0$$

Since $-k$ and k have exactly the same eigenfunctions we see that the multiplicity of each eigenvalue λ_j is

$$(n-1)\cdot\#\{k \mid k \in Z^n,\ |k|^2 = \frac{L^2}{4\pi^2}\lambda_j = \frac{\lambda_j}{\lambda_1}\}.$$

Proposition 4.14. The asymptotic behavior of the eigenvalues λ_j is given by

(4.43) $$\lim_{j\to\infty} j^{-2/n}(\frac{\lambda_j}{\lambda_1}) = ((n-1)\omega_n)^{-2/n} .$$

Proof. Let $N_\lambda = \#\{m \mid \lambda_m \le \lambda\}$. Clearly,

$$N_\lambda = (n-1)\cdot\#\{k \in Z^n \setminus \{0\} \mid |k| \le \sqrt{\frac{\lambda}{\lambda_1}}\}$$

For each k with integer coordinates there will be a box of unit volume that contains no other point with integer coordinates: $B_k = k + (-1/2,1/2)^n$. This box is included in a ball of radius $\sqrt{n}/2$ around k. Thus

$$\{x \mid |x| < \sqrt{\frac{\lambda}{\lambda_1}} - \frac{\sqrt{n}}{2}\} \subset \bigcup_{|k| \le \sqrt{\lambda/\lambda_1}} B_k \quad \{x \mid |x| \le \sqrt{\frac{\lambda}{\lambda_1}} + \frac{\sqrt{n}}{2}\}.$$

The first inclusion follows from the fact that the distance from any point $x \in \mathbb{R}^n$ to Z^n is at most $\sqrt{n}/2$. Since the volume of $\bigcup_{|k| \le \sqrt{\lambda/\lambda_1}} B_k = \frac{N_\lambda}{n-1} + 1$ it follows that

$$\omega_n\left(\sqrt{\frac{\lambda}{\lambda_1}} - \frac{\sqrt{n}}{2}\right)^n \le \frac{N_\lambda}{n-1} + 1 \le \omega_n\left(\sqrt{\frac{\lambda}{\lambda_1}} + \frac{\sqrt{n}}{2}\right)^n.$$

Now, if $\lambda' < \lambda_j \le \lambda''$ then $N(\lambda') < j \le N(\lambda'')$ and thus

$$(n-1)\left(\left(\sqrt{\frac{\lambda'}{\lambda_1}} - \frac{\sqrt{n}}{2}\right)^n \omega_n - 1\right) \le j \le (n-1)\left[\omega_n\left(\sqrt{\frac{\lambda''}{\lambda_1}} + \frac{\sqrt{n}}{2}\right)^n - 1\right]$$

Allowing $\lambda' \to \lambda_j$, $\lambda'' \to \lambda_j$ we get

$$(n-1)\left[\omega_n\left(\sqrt{\frac{\lambda_j}{\lambda_1}} - \frac{\sqrt{n}}{2}\right)^n - 1\right] \le j \le (n-1)\left[\omega_n\left(\sqrt{\frac{\lambda_j}{\lambda_1}} + \frac{\sqrt{n}}{2}\right)^n - 1\right].$$

From this inequality it follows that

$$\text{(4.44)} \qquad \frac{\lambda_j}{\lambda_1} = ((n-1)\omega_n)^{-2/n} j^{2/n} + O(j^{1/n}).$$

The result of Proposition 4.14 and actually even (4.44) remain true for the Stokes operator in a bounded C^∞ domain Ω ([Ko 1]) provided we replace λ_1 by its large volume asymptotic value, $|\Omega|^{-2/n}(2\pi)^2$. Of course, in the periodic case $\lambda_1 = |Q_L|^{-2/n}(2\pi)^2 = \frac{4\pi^2}{L^2}$.

5

THE NAVIER-STOKES EQUATIONS

Let $\Omega \subset \mathbb{R}^n$ be an open set. The Navier-Stokes equations are a n+1 by n+1 system of equations for the unknowns $u_1(t,x),\ldots,u_n(t,x)$ representing a velocity vector and $p(t,x)$ representing specific pressure. The variables $(t,x) \in \mathbb{R}_+ \times \Omega$ represent time and position. The equations are

$$(5.1) \qquad \frac{\partial u_i}{\partial t} - \nu\Delta u_i + u_j \frac{\partial u_i}{\partial x_j} + \frac{\partial p}{\partial x_i} = f_i, \quad i = 1,\ldots,n,$$

$$(5.2) \qquad \text{div } u = \frac{\partial u_i}{\partial x_i} = 0.$$

The vector $(u_j \frac{\partial u_i}{\partial x_j})$ $i = 1,\ldots,n$ is denoted $(u\cdot\nabla)u$ or $u\cdot\nabla u$. The functions $f_i(t,x)$ are given specific body forces. The coefficient $\nu > 0$ is called the kinematic viscosity coefficient. The equations are supplemented by boundary conditions. We shall treat "no-slip" boundary conditions

$$(5.3) \qquad u(t,x) = 0 \quad \text{for } x \in \partial\Omega$$

or periodic boundary conditions

$$(5.3') \qquad u(t,x+Le_i) = u(t,x), \quad L > 0, \ e_i \text{ the canonical basis in } \mathbb{R}^n.$$

The initial value problem consists in solving (5.1)-(5.3) together with the initial condition

(5.4) $u(0,x) = u_0(x)$

where u_0 is a given vector function.

The Navier-Stokes equations possess important scaling properties. Suppose that the functions $v(s,y)$, $q(s,y)$ solve the system

$$\begin{cases} \frac{\partial v_i}{\partial s} - \mu\Delta v_i + v_j \frac{\partial v_i}{\partial y_j} + \frac{\partial q}{\partial y_i} = g_i(s,y) & i = 1,\dots,n \\ \frac{\partial v_i}{\partial y_i} = 0 \end{cases}$$

for $s > 0$, $y \in D \subset \mathbb{R}^n$. Then for $L > 0$, $T > 0$ the functions

(5.5) $u(t,x) = \frac{L}{T} v(\frac{t}{T}, \frac{x}{L})$

(5.6) $p(t,x) = \frac{L^2}{T^2} q(\frac{t}{T}, \frac{x}{L})$

solve the equations

$$\begin{cases} \frac{\partial u}{\partial t} - \nu\Delta u + (u\cdot\nabla)u + \nabla p = f \\ \operatorname{div} u = 0 \end{cases}$$

in $\Omega = LD$ for $t > 0$ with

(5.7) $f = \frac{L}{T^2} g(\frac{t}{T}, \frac{x}{L})$

and

(5.8) $\nu = \frac{L^2}{T} \mu$.

We refer to (5.5)-(5.6) as the scaling properties of the Navier-Stokes equations. All statements about Navier-Stokes equations must be made bearing these scaling properties in mind. We say that u has the scaling dimension L/T, p has the dimension L^2/T^2, f the dimension L/T^2 and ν the dimension L^2/T. The dimension of the variable t is T and that of x is L.

Applying Leray's projector P to (5.1) we obtain for smooth functions u(t,x), p(t,x) satisfying (5.1), (5.2) that

$$(5.9) \qquad \frac{du}{dt} + \nu Au + B(u,u) = Pf$$

where A is the Stokes operator and

$$(5.10) \qquad B(u,u) = P(u \cdot \nabla u).$$

The procedure of applying P eliminates the pressure from the equations. In (5.9) the term νAu is the dissipative term. Its effect is one of dissipating energy and smoothing. The term B(u,u) is the nonlinear term and Pf is the forcing term.

6

INEQUALITIES FOR THE NONLINEAR TERM

We first recall the Sobolev imbedding inequalities.

If Ω is an open set in $\mathbb{R}^n$ of class C^1 then for $m < n/2$ one has $H^m(\Omega) \subset L^q(\Omega)$ where

$$(6.1) \qquad \frac{1}{q} = \frac{1}{2} - \frac{m}{n}$$

with continuous inclusion

$$(6.2) \qquad \|u\|_{L^q(\Omega)} \leq c_m \|u\|_{m,\Omega}.$$

Note that (6.1) is the only possible definition of a q that makes (6.2) scale invariant. If Ω is bounded and is of class C^ℓ then for $\ell \geq m > n/2$, $H^m(\Omega) \subset C^k(\overline{\Omega})$ where $k = m - [\frac{n}{2}] - 1 \geq 0$ and

$$(6.3) \qquad \| D^\alpha u \|_{L^\infty(\Omega)} \leq C_{\alpha,m} |\Omega|^{\frac{1}{n}(m - \frac{n}{2} - |\alpha|)} \|u\|_{m,\Omega}$$

for all $|\alpha| \leq k$.

In the case $\Omega = \mathbb{R}^n$ the spaces H^s can be easily defined, via Fourier transform, for noninteger values of s and we have

$$(6.4) \qquad \|u\|_{L^q(\mathbb{R}^n)} \leq c_s \|u\|_{H^s(\mathbb{R}^n)}, \quad s < \frac{n}{2}, \ \frac{1}{q} = \frac{1}{2} - \frac{s}{n}$$

$$(6.5) \qquad \| D^\alpha u \|_{L^\infty(\mathbb{R}^n)} \leq c_{\alpha,s} \|u\|_{H^s(\mathbb{R}^n)}, \ s > n/2, \quad |\alpha| < s - \frac{n}{2}$$

In section 4 we defined the extension maps $E^\ell : H^\ell(\Omega) \to H^\ell(\mathbb{R}^n)$. Using them we can introduce noninteger valued Sobolev morms on Ω. Let us note that $E^{\ell}|_{H^{\ell+1}(\Omega)} = i_{\ell}E^{\ell+1}$ where $i_{\ell}: H^{\ell+1}(\mathbb{R}^n) \hookrightarrow H^{\ell}(\mathbb{R}^n)$ is the inclusion. Let $u \in H^{\ell+1}(\Omega)$, Ω of class $C^{\ell+1}$. Then one can take for $s = (1-t)\ell + t(\ell+1)$, $t \in (0,1)$

(6.6) $\quad \|u\|_{s,\Omega} = \| E^{\ell+1}u\|_{s,R^n}$

We note that, for every $u \in H^{\ell+1}(\Omega)$, $\ell < s < \ell+1$

(6.7) $\quad \|u\|_{s,\Omega} \leq c_{\ell}\|u\|^{1-t}_{\ell,\Omega}\|u\|^{t}_{\ell+1,\Omega}$

with $t = s - \ell$. Indeed

$$\| E^{\ell+1}u \|_{s,R^n} \leq \| E^{\ell+1}u \|^{1-t}_{\ell} \| E^{\ell+1}u \|^{t}_{\ell+1}$$

and the properties of $E^{\ell+1}$ assure $\| E^{\ell+1}u \|_{k,R^n} \leq c_k\|u\|_{k,\Omega}$ for $k \leq \ell+1$. We shall avoid using the spaces obtained by completing $H^{\ell+1}(\Omega)$ with the norms (6.6) and state our inequalities in terms of the right-hand side of (6.7) when, in fact, the left-hand side would do.

Let us define first a trilinear form by

(6.8) $\quad b(u,v,w) = \int_{\Omega} u_j \frac{\partial v_i}{\partial x_j} w_i dx = \int_{\Omega} \langle u\cdot\nabla v, w\rangle dx.$

We assume Ω is bounded and of class C^{ℓ} with ℓ sufficiently large. The expression (6.8) surely makes sense for functions $u,v,w \in C^{\infty}(\bar{\Omega})^n$.

Proposition 6.1. Let $\Omega \subset \mathbb{R}^n$ be bounded, open and of class C^{ℓ}. Let s_1,s_2,s_3 be real numbers, $0 \leq s_1 \leq \ell$, $0 \leq s_2 \leq \ell-1$, $0 \leq s_3 \leq \ell$. Let us assume that

(i) $\quad s_1 + s_2 + s_3 \geq n/2$ if $s_i \neq n/2$ for all $i = 1,2,3$

or

(ii) $\quad s_1 + s_2 + s_3 > n/2$ if $s_i = n/2$ for at least one i.

(In other words, we assume $s_1 + s_2 + s_3 \geq n/2$ and $(s_1,s_2,s_3) \neq (0,0,n/2)$, $(0,n/2,0)$, $(n/2,0,0)$).

Then there exists a constant depending on s_1,s_2,s_3,Ω, scale invariant, such that

$$(6.9)\qquad |b(u,v,w)| \leq c|\Omega|^{\frac{s_1+s_2+s_3}{n}-\frac{1}{2}} \|u\|_{[s_1],\Omega}^{1+[s_1]-s_1} \|u\|_{[s_1]+1,\Omega}^{s_1-[s_1]} \cdot$$

$$\cdot \|v\|_{[s_2]+1,\Omega}^{1+[s_2]-s_2} \|v\|_{[s_2]+2,\Omega}^{s_2-[s_2]} \|w\|_{[s_3],\Omega}^{1+[s_3]-s_3} \|w\|_{[s_3]+1,\Omega}^{s_3-[s_3]}$$

for all $u,v,w \in C^\infty(\overline{\Omega})^n$.

Remark. The estimate (6.9) is the extrapolated version of

$$(6.10)\qquad |b(u,v,w)| \leq c|\Omega|^{\frac{s_1+s_2+s_3}{n}-\frac{1}{2}} \|u\|_{s_1,\Omega}\|v\|_{s_2+1,\Omega}\|w\|_{s_3,\Omega}$$

Proof. Let us set $\tilde{u} = E^\ell u$, $\tilde{v} = E^\ell v$, $\tilde{w} = E^\ell w$. Let us consider first the case

$s_i < n/2$ for all $i = 1,2,3$. Defining q_i by $\frac{1}{q_i} = \frac{1}{2} - \frac{s_i}{n}$ $i = 1,2,3$ and q_4 $(1 \leq q_4 \leq \infty)$ by $\frac{1}{q_1} + \frac{1}{q_2} + \frac{1}{q_3} + \frac{1}{q_4} = 1$ which is possible since $\sum_{i=1}^{3} \frac{1}{q_i} = \frac{3}{2} - \frac{\sum s_i}{n} \leq 1$ we have, by Hölder's inequality

$$\left|\int_\Omega u_j \frac{\partial v_i}{\partial x_j} w_i \, dx\right| = \left|\int_\Omega u_j \frac{\partial v_i}{\partial x_j} w_i \cdot 1 \, dx\right|$$

$$\leq \|u\|_{L^{q_1}(\Omega)} \|\nabla v\|_{L^{q_2}(\Omega)} \|w\|_{L^{q_3}(\Omega)} \|1\|_{L^{q_4}(\Omega)}$$

$$\leq |\Omega|^{\frac{\sum_{i=1}^{3} s_i}{n}-\frac{1}{2}} \|\tilde{u}\|_{L^{q_1}(\mathbb{R}^n)} \|\nabla\tilde{v}\|_{L^{q_2}(\mathbb{R}^n)} \|\tilde{w}\|_{L^{q_3}(\mathbb{R}^n)}$$

$$\leq c|\Omega|^{\frac{\sum_1^3 s_i}{n}-\frac{1}{2}} \|\tilde{u}\|_{s_1} \|\nabla\tilde{v}\|_{s_2} \|\tilde{w}\|_{s_3}$$

This proves (6.10) in this case.

If one of the s_i -s is larger than n/2 we replace the corresponding q_i by ∞ and use (6.5) for $\alpha = 0$. This would prove

$$|b(u,v,w)| \leq k\| \tilde{u} \|_{s_1}\| \nabla\tilde{v} \|_{s_2}\| \tilde{w} \|_{s_3}$$

with k depending on Ω, s_1, s_2, s_3 but not scale invariant. The reason is that (6.5) is not dilation invariant. In order to obtain a scale invariant estimate we use Lemma 4.9 because (4.19) is dilation invariant. Suppose for example $s_1 > n/2$, $s_2 < n/2$, $s_3 < n/2$. Then write $n/2 = (1-t)0 + ts_1$ for $t \in (0,1)$. We use

$$\| \tilde{u} \|_{L^\infty(\mathbb{R}^n)} \leq \| \tilde{u} \|^{1-t}_{L^2(\mathbb{R}^n)}\| \tilde{u} \|^{t}_{s_1,\mathbb{R}^n} \quad \text{(see 4.19).}$$

Then use the extrapolation, since $\| \tilde{u} \|_{L^2(\mathbb{R}^n)} \leq d\|u\|_{0,\Omega}$:

$$(6.11) \quad \|u\|_{L^\infty(\Omega)} \leq d\|u\|^{1-t}_{L^2(\Omega)}\left[\|u\|^{1+[s_1]-s_1}_{[s_1],\Omega} \|u\|^{s_1-[s_1]}_{[s_1]+1,\Omega}\right]^{t}$$

and also

$$\|u\|_{L^2(\Omega)} \leq |\Omega|^{\frac{[s_1]}{n}} \|u\|_{[s_1],\Omega} ,$$

$$\|u\|_{L^2(\Omega)} \leq |\Omega|^{\frac{[s_1]+1}{n}} \|u\|_{[s_1]+1,\Omega}$$

(see the definition (4.13) of $\|u\|_{m,\Omega}$). We obtain, because $(1 + [s_1] - s_1) + s_1 - [s_1] = 1$

$$(6.12) \quad \|u\|_{L^2(\Omega)} = \|u\|^{1+[s_1]-s_1}_{L^2(\Omega)} \|u\|^{s_1-[s_1]}_{L^2(\Omega)}$$

$$\leq |\Omega|^{\frac{[s_1]}{n}\cdot(1+[s_1]-s_1) + \frac{[s_1]+1}{n}(s_1-[s_1])} \|u\|^{1+[s_1]-s_1}_{[s_1],\Omega} \|u\|^{s_1-[s_1]}_{[s_1]+1,\Omega}$$

Then, from (6.11) and (6.12) we get

$$(6.13)\qquad \|u\|_{L^\infty(\Omega)} \le c|\Omega|^{\frac{s_1}{n}-\frac{1}{2}} \|u\|_{[s_1],\Omega}^{1+[s_1]-s_1} \|u\|_{[s_1]+1,\Omega}^{s_1-[s_1]}.$$

Indeed $t = n/2s_1$ and

$$\frac{[s_1]}{n}(1+[s_1]-s_1) + \frac{[s_1]+1}{n}(s_1-[s_1]) = \frac{s_1}{n}.$$

In this case, using (6.13), q_2, q_3 defined as usual and $\frac{1}{q_2}+\frac{1}{q_3}+\frac{1}{q_4} = 1$ defining q_4 we get (6.9).

The case in which more than one of the s_i is larger than n/2 is treated in the same way. If one or more of the $s_i = n/2$ and $s_1 + s_2 + s_3 > n/2$ we can modify s_i to s_i', $s_i' \le s_i$, $s_1' + s_2' + s_3' > n/2$, $s_i' \ne n/2$ all i = 1,2,3. We write then (6.9) for the s_i' and note that it implies the one for s_i because

$$\|f\|_{m',\Omega} \le \|f\|_{m,\Omega} |\Omega|^{\frac{m-m'}{n}} \qquad \text{if } m \ge m'.$$

Let us note that (6.9) permits us to define b(u,v,w) for u,v,w such that the right-hand side of (6.9) is finite.

Remark 6.2. The case in which (s_1, s_2, s_3) is one of the vectors (n/2,0,0), (0,n/2,0) or (0,0,n/2) corresponds to an estimate of the type L^∞ (for the two remaining places) Combining with Lemma 4.10 we get estimates of the type

$$(6.14)\qquad |b(u,v,w)| \le c\|u\|_{L^\infty}\|v\|_{1,\Omega}\|w\|_{0,\Omega} \le c\|u\|_{\ell',\Omega}^{1-t}\|u\|_{\ell,\Omega}^{t}\|v\|_{1,\Omega}\|w\|_{0,\Omega}$$

for $n/2 = (1-t)\ell' + t\ell$, t (0,1), Ω of class C^ℓ at least

$$(6.15)\qquad |b(u,v,w)| \le c\|u\|_{0,\Omega}\|v\|_{1,\Omega}\|w\|_{L^\infty} \le c\|u\|_{0,\Omega}\|v\|_{1,\Omega}\|w\|_{\ell',\Omega}^{1-t}\|w\|^{t}$$

and

(6.16) $|b(u,v,w)| \leq c\|u\|_{0,\Omega}\|w\|_{0,\Omega}\|\nabla v\|_{L^\infty} \leq c\|u\|_{0,\Omega}\|w\|_{0,\Omega}\|v\|^{1-t}_{\ell'+1,\Omega}\|v\|^{t}_{\ell+1,\Omega}$

Until now we did not assume that any of u,v or w is divergence free. Clearly, if div u = 0 and $<\gamma_0 w \cdot \gamma_0 v> \cdot \gamma u = 0$ (i.e., $<w,v> <u.n_\Omega> = 0$ on $\partial\Omega$, n_Ω being the normal) then

$$\int_\Omega u_j \frac{\partial v_i}{\partial x_j} w_i = - \int u_j \frac{\partial w_i}{\partial x_j} v_i$$

In particular, if $u,v,w \in \mathcal{V}$

(6.17) $b(u,v,w) = -b(u,w,v)$

(6.18) $b(u,v,v) = 0$

The properties (6.17), (6.18) hold of course for $u,v,w \in V$.

In Proposition 4.12 we proved that, as long as $0 \leq m \leq \ell$ and Ω is bounded and of class C^ℓ, $\mathcal{D}(A^{m/2}) \subset H^m(\Omega)^n$. We can extend this result to noninteger values of m. The proof of this statement lies outside the scope of this work; we shall give however an indication of its flavor.

First we remark that $\mathcal{D}(A^{\alpha/2})$ is the interpolation space $[\mathcal{D}(A^{[\alpha]/2}), \mathcal{D}(A^{\frac{[\alpha]+1}{2}})]_{\alpha-[\alpha]}$, [X,Y] denoting complex interpolation. (See [Be-Lo]). This follows from the fact that $\mathcal{D}(A^{\beta/2})$ are isomorphic to spaces $\ell^{2,\beta}:\{(u_j)| \sum \lambda_j^\beta u_j^2 < \infty\}$ through the obvious map $u \mapsto ((u,w_j)_j)$ and from the fact that the interpolation space $[\ell^{2,m}, \ell^{2,m'}]_\theta = \ell^{2,(1-\theta)m+\theta m'}$. Then we observe that the definition

$$H^\alpha(\Omega)^n = \{u \in L^2(\Omega)^n \mid E^\ell u \in H^\alpha(\mathbb{R}^n)^n\}$$

where $E^\ell: H^\ell(\Omega)^n \to H^\ell(\mathbb{R}^n)$ is our extension map (recall $E^\ell: H^m(\Omega)^n \to H^m(\mathbb{R}^n)^n$, $0 \leq m \leq \ell$ is bounded) agrees with

$$H^\alpha(\Omega)^n = [H^{[\alpha]}(\Omega)^n, H^{[\alpha]+1}(\Omega)^n]_{\alpha-[\alpha]} .$$

This can be proven as a consequence of the fact that $H^\alpha(R^n)^n = [H^{[\alpha]}(R^n)^n, H^{[\alpha]+1}(R^n)^n]_{\alpha-[\alpha]}$. Indeed if $[\alpha]+1 \le \ell$ then, from the interpolation theorem it follows that

$$E^\ell : [H^{[\alpha]}(\Omega)^n, H^{[\alpha]+1}(\Omega)^n]_{\alpha-[\alpha]} \to H^\alpha(R^n)^n$$

is bounded and thus

$$[H^{[\alpha]}(\Omega)^n, H^{[\alpha]+1}(\Omega)^n]_{\alpha-[\alpha]} \subset H^\alpha(\Omega)^n$$

by the definition of $H^\alpha(\Omega)^n = \{u \in L^2(\Omega)^n | E^\ell u \in H^\alpha(R^n)\}$. The other inclusion follows using the interpolation theorem for the (obviously bounded) restriction operator $R: H^{[\alpha]+1}(R^n) \to H^{[\alpha]+1}(\Omega)^n$, $R: H^{[\alpha]}(R^n)^n \to H^{[\alpha]}(\Omega)^n$. This operator is then bounded $R: H^\alpha(R^n)^n \to [(H^{[\alpha]}(\Omega))^n (H^{[\alpha]+1}(\Omega))^n]_{\alpha-[\alpha]}$ and therefore, if $u \in H^\alpha(\Omega)^n$ then by definition $E^\ell(u) \in H^\alpha(R^n)$ and $u = R(E^\ell u)$ belongs to $[H^{[\alpha]}(\Omega), H^{[\alpha]+1}(\Omega)]_{\alpha-[\alpha]}$. Now, the inclusion map $i: \mathcal{D}(A^{\frac{m+1}{2}}) \hookrightarrow H^{m+1}(\Omega)^n$ and $i: \mathcal{D}(A^{m/2}) \hookrightarrow H^m(\Omega)^n$ is bounded if $1+m \le \ell$ by Proposition 4.12. If $[\alpha] = m$, $[\alpha]+1 = m+1$ if follows that i is bounded from

$$i: [\mathcal{D}(A^{\frac{m}{2}}), \mathcal{D}(A^{\frac{m+1}{2}})]_{\alpha-m} \hookrightarrow [H^m(\Omega)^n, H^{m+1}(\Omega)^n]_{\alpha-m}$$

i.e., that $\mathcal{D}(A^{\alpha/2}) \hookrightarrow H^\alpha(\Omega)$ for all α such that $[\alpha]+1 \le \ell$.

<u>Definition 6.3</u>. Let $u,v \in C^\infty(\overline{\Omega})^n$. We define $B(u,v)$ by

(6.19) $\quad B(u,v) = P(u\cdot\nabla v)$

where P is Leray's projector.

Proposition 6.4. Let $\Omega \subset \mathbb{R}^n$ be bounded, open, of class C^ℓ. Let s_1, s_2, s_3 be real numbers such that $0 \le s_1 \le \ell$, $0 \le s_2 \le \ell-1$, $0 \le s_3 \le \ell$. Assume that $s_1 + s_2 + s_3 \ge n/2$ and that (s_1, s_2, s_3) is neither of the triplets $(n/2,0,0)$, $(0,n/2,0)$, $(0,0,n/2)$. Then there exists a constant $c = c(s_1,s_2,s_3,\Omega)$ scale invariant, such that, for all $u,v \in C^\infty(\overline{\Omega})^n$,

$$(6.20) \qquad |A^{-\frac{s_3}{2}} B(u,v)| \le c|\Omega|^{\frac{s_1+s_2+s_3}{n} - \frac{1}{2}} \|u\|_{[s_1],\Omega}^{1+[s_1]-s_1} \|u\|_{[s_1]+1,\Omega}^{s_1-[s_1]}$$

$$\cdot \|v\|_{[s_2]+1,\Omega}^{1+[s_2]-s_2} \|v\|_{[s_2]+2,\Omega}^{s_2-[s_2]}$$

Remark. The right-hand side of (6.20) can be replaced by

$$c|\Omega|^{\frac{s_1+s_2+s_3}{n} - \frac{1}{2}} \|u\|_{s_1,\Omega} \|u\|_{s_2+1,\Omega}$$

where $\|u\|_{s,\Omega} = \| E^\ell u \|_{s,R^n}$.

Proof of Proposition 6.4. Clearly, since $B(u,v)$ is an element of H, $A^{-s_3/2} B(u,v)$ belongs to $\mathcal{D}(A^{s_3/2})$. Let $w \in H$ be arbitrary. Then

$$(A^{-s_3/2} B(u,v),w) = (B(u,v),A^{-s_3/2} w).$$

Since $P(A^{-s_3/2} w) = A^{-s_3/2} w$ it follows that

$$(A^{-s_3/2} B(u,v),w) = b(u,v,A^{-s_3/2} w).$$

Now by (6.10)

$$|b(u,v,A^{-s_3/2} w|$$
$$\le c|\Omega|^{\frac{s_1+s_2+s_3}{n} - \frac{1}{2}} \|u\|_{s_1,\Omega} \|v\|_{s_2+1,\Omega} \| A^{-s_3/2} w \|_{s_3,\Omega}.$$

From Proposition 4.12, if s_3 is an integer and from the discussion preceding Definition 6.3, it follows that $A^{-s_3/2}w \in \mathcal{D}(A^{s_3/2}) \subset H^{s_3}(\Omega)^n$ and

$$\| A^{-s_3/2} w \|_{s_3,\Omega} \leq \| A^{s_3/2} A^{-s_3/2} w \|_0 = \| w \|_{0,\Omega}.$$

Thus

$$|(A^{-s_3/2} B(u,v),w)| \leq c|\Omega|^{\frac{s_1+s_2+s_3}{n} - \frac{1}{2}} \|u\|_{s_1,\Omega} \|v\|_{s_{2+1},\Omega} \|w\|_{0,\Omega}.$$

7

STATIONARY SOLUTIONS TO THE NAVIER-STOKES EQUATIONS

Stationary (steady state, equilibrium) solutions of the Navier-Stokes equations are time independent solutions:

$$(7.1) \qquad \nu Au + B(u,u) = f, \quad u \in \mathcal{D}(A)$$

for $f \in H$, given.

In this section $\Omega \subset \mathbb{R}^n$ $n = 2$ or 3, and Ω is of class C^ℓ, $\ell \geq 2$. If $u \in \mathcal{D}(A)$ then $B(u,u) \in H$ so (7.1) is an equation in H. Moreover, if $u \in \mathcal{D}(A)$ then (7.1) can be written as the Stokes system

$$-\nu\Delta u + \text{grad } p = f - (u \cdot \nabla)u$$

$$\text{div } u = 0$$

$$u|_{\partial\Omega} = 0$$

and we see that $p \in H^1(\Omega)$.

We discuss first existence solutions to (7.1) in $\mathcal{D}(A)$. We call them simply solutions or strong solutions. As in the case of the Stokes system, we can use a variational approach and define weak solutions. By this we mean solutions $u \in V$ of the equation

$$(7.2) \qquad \nu((u,v)) + b(u,u,v) = (f,v), \quad \text{for all } v \in V.$$

This formulation (7.2) makes sense for $f \in V'$.

Lemma 7.1. Let $u \in V$. Let us define the map $K(u):V \to V$ by $K(u)v = A^{-1}B(u,v)$. Then the map $K(u):V \to V$ is compact. More precisely, $K(u):V \to \mathcal{D}(A^{3/4})$ is bounded and

$$(7.3) \qquad |A^{3/4}K(u)v| \leq c\|u\|\,\|v\| \qquad \text{for some } c > 0.$$

Proof. $A^{3/4}K(u)v = A^{3/4}A^{-1}B(u,v) = A^{-1/4}B(u,v)$. The inequality (7.3) follows thus from (6.20) with $s_1 = 1$, $s_2 = 0$, $s_3 = 1/2$, and $\frac{3}{2} \geq \frac{n}{2}$ for $n = 2$ or 3. We omit the dependence of the constant on Ω and do not write (7.3) in scale invariant form. Compactness follows from the fact that $\mathcal{D}(A^{3/4}) \subset H^{3/2}(\Omega)^n$ and the inclusion $V \cap (H^{3/2}(\Omega))^n \hookrightarrow V$ is compact by Rellich's selection theorem.

Lemma 7.2. Let B be a closed ball in $\mathbb{R}^n$. Suppose $\phi:B \to \mathbb{R}^n$ satisfies: ϕ continuous and $\langle\phi(v),v\rangle < 0$ for all $v \in \partial B$. Then there exists $v \in B$ such that $\phi(v) = 0$.

Proof. Let us assume that R is the radius of the ball. Assume that $\phi(v) \neq 0$ for all $v \in B$. Then the map $v \mapsto R\,\frac{\phi(v)}{|\phi(v)|}$ is a map from B into ∂B. As a continuous map from B to B it has a fixed point by Brouwer's theorem $v_0 = R\,\frac{\phi(v_0)}{|\phi(v_0)|}$. Moreover $v_0 \in \partial B$. But this is absurd since

$$\langle v_0,v_0\rangle = \frac{\langle\phi(v_0),v_0\rangle}{|\phi(v_0)|}R < 0.$$

Remark. This proves that the system $\frac{dx}{dt} = \phi(x(t))$ in $\mathbb{R}^n$ has a fixed point $x_0 \in B$.

Let us introduce some notation. Let m be an integer. We denote by P_m the projection in H on the span of $w_1, w_2, \ldots, w_m$, the first m eigenfunctions of A. In this section we consider also the corresponding object in V and denote it by P_m^V. Since (w_k) are orthogonal in

$V = \mathcal{D}(A^{1/2})$ we have the formulae

(7.4) $$P_m f = \sum_{j=1}^{m} (w_j,f)w_j , \qquad f \in H$$

(7.5) $$P_m^V f = \sum_{j=1}^{m} \lambda_j^{-1}((w_j,f))w_j , \quad f \in V$$

(Remark: $P_m f = P_m^V f$ for $f \in V$).

Theorem 7.3. Let $f \in H$. There exists a solution $u \in \mathcal{D}(A)$ of (7.1).

Proof. Let $m > 0$ be an integer. We first remark that (7.1) is equivalent to

(7.6) $$\nu u + K(u)u = A^{-1}f = g.$$

We shall approximate (7.6) by

(7.7) $$\nu u + P_m^V K(u)u = P_m^V(A^{-1}f).$$

The first step is to prove that for every $m > 0$ integer, there exists a solution $u_m \in P_m^V(V)$ of (7.7).

Let $-\Phi_m(u) = \nu u + P_m^V(K(u)u) - P_m^V(g)$, for $u \in P_m^V(V)$. Then

$$-((\Phi_m(u),u)) = ((\nu u,u)) + ((K(u)u,\ P_m^V u)) - ((P_m^V g,u))$$

$$= \nu\|u\|^2 + ((K(u)u,u)) - ((g,u)).$$

Now $(B(u,v),v) = 0$ for all $v \in V$ means

(7.8) $$((K(u)v,v)) = 0 \quad \text{for all } v \quad V.$$

So $((\Phi_m(u),u)) = -\nu\|u\|^2 + ((g,u))$. Therefore, according to Lemma 7.2 there exists $u = u_m$ such that $\Phi_m(u) = 0$ provided we choose the ball in P_m^V of radius $R > \frac{\|g\|}{\nu}$.

We obtain thus a solution $u_m \in P_m^V(V)$ of (7.7). Moreover

$$(7.9) \qquad \|u_m\| \le \frac{\|A^{-1}f\|}{\nu} = \frac{|A^{-1/2}f|}{\nu}.$$

Now, using (7.3) it follows that

$$\nu|A^{3/4}u_m| \le |A^{3/4}P_m^V K(u_m)u_m| + |A^{-1/4}f| = |P_m^V A^{3/4}K(u_m)u_m| + |A^{-1/4}f|$$

$$\le c\|u_m\|^2 + |A^{-1/4}f| + c\,\frac{|A^{-1/2}f|^2}{\nu^2} + |A^{-1/4}f|.$$

We used implicitly the fact that $u_m \in \mathcal{D}(A)$. But the space $P_m^V V = P_m H$ is the span of $\{w_1,\dots,w_m\}$ and thus is contained in $\mathcal{D}(A^\alpha)$ for all $\alpha > 0$. We have thus

$$(7.10) \qquad |A^{3/4}u_m| \le c\,\frac{|A^{-1/2}f|^2}{\nu^3} + \frac{|A^{-1/4}f|}{\nu}$$

Now from (7.9) it follows that the sequence u_m is bounded in V. There exists a subsequence $u_{m'}$ which is weakly convergent in V to some element u. We could use the information given by (7.10) and improve the convergence but we refrain from doing so, for the moment.

Passing to a subsequence if necessary, we may assume that $u_{m'} \to u$ in H strongly. Then we claim that $P_{m'}^V K(u_{m'})u_{m'}$ converges to $K(u)u$ in V, weakly. Indeed, this statement is equivalent to $A^{1/2}P_{m'}^V K(u_{m'})u_{m'}$ converges to $A^{1/2}K(u)u$ in H, weakly. Let $h \in H$ be arbitrary. Then

$$(A^{1/2}P_{m'}^V K(u_{m'})u_{m'} - A^{1/2}K(u)u,h) = b(u_{m'},u_{m'},P_{m'}A^{-1/2}h) - b(u,u,A^{-1/2}h)$$

$$= b(u_{m'} - u,u_{m'},A^{-1/2}h) + b(u,u_{m'} - u,A^{-1/2}h) + b(u_{m'},u_{m'},(I - P_{m'})A^{-1/2}h)$$

$$= b(u_{m'} - u,u_{m'},A^{-1/2}h) - b(u_{m'},A^{-1/2}h,u_{m'} - u) + b(u_{m'},u_{m'},(I-P_{m'})A^{-1/2}).$$

Now we use estimate (7.9) in conjunction with (6.9) with $s_1 = 1/2$, $s_2 = 0$, $s_3 = 1$ for the first term in the sum, $s_1 = 1$, $s_2 = 0$, $s_3 = 1/2$ for the second and third terms in the sum

$$|(A^{1/2}P^V_{m'}K(u_{m'}),u_{m'} - A^{1/2}K(u)u,h)|$$

$$\leq c\frac{|A^{-1/2}f|}{\nu}\left(\frac{|A^{-1/2}f|}{\nu} + \|u\|\right)^{1/2}|u_{m'}-u|^{1/2}|h|$$

$$+ \left(\frac{|A^{-1/2}f|}{\nu}\right)^2|h|^{1/2}|(1 - P_{m'})A^{-1/2}h|^{1/2} .$$

Now, since $|A^{-1/2}(1 - P_{m'})h|^{1/2} \leq \lambda_{m'+1}^{-1/4}|h|^{1/2}$ it follows that indeed $P^V_{m'}K(u_{m'},u_{m'})$ converges weakly in V to K(u)u.

Since $P^V_{m'}(A^{-1}f)$ converges (strongly even) in V to $A^{-1}f$ it follows, passing to the (weak) limit in V in (7.7) that $u \in V$ solves (7.6). Then using (7.10) and passing to lim sup we deduce that $u \in \mathcal{D}(A^{3/4})$ and

$$(7.11) \qquad |A^{3/4}u| \leq c\,\frac{|A^{-1/2}f|^2}{\nu^3} + \frac{|A^{-1/4}f|}{\nu}$$

Now $AK(u)u = B(u,u)$ and using (6.9) with $s_1 = 1$, $s_2 = 1/2$ and (7.11) it follows that

$$|AK(u)u| \leq c\|u\||A^{3/4}u| \leq c\,\frac{|A^{-1/2}f|}{\nu}\left[\frac{|A^{-1/2}f|^2}{\nu^3} + \frac{|A^{-1/2}f|}{\nu}\right].$$

Then since $u = -\frac{1}{\nu}K(u)u + \frac{A^{-1}f}{\nu}$ it follows that $u \in \mathcal{D}(A)$ and that

$$(7.12) \qquad |Au| \leq \frac{|f|}{\nu} + c\left[\frac{|A^{-1/2}f|^3}{\nu^5} + \frac{|A^{-1/4}f||A^{-1/2}f|}{\nu^2}\right]$$

This concludes the proof of the theorem.

Remark. (1) If f V', i.e., if $A^{-1/2}f \in H$ the same proof provides a solution to (7.2) which satisfies $\|u\| \leq \frac{|A^{-1/2}f|}{\nu}$.

(2) If the boundary of Ω is of class C^{ℓ}, $\ell \geq 2$, and if $f \in H^{\alpha}(\Omega)^n$, $\alpha \leq \ell$ then $u \in H^{\alpha+2}(\Omega)^n$.

8

WEAK SOLUTIONS OF THE NAVIER-STOKES EQUATIONS

We consider $\Omega \subset \mathbb{R}^d$, d = 2 or 3 a bounded open set of class C^ℓ, $\ell \geq 2$ large enough. The Navier-Stokes equations will be written in the form

(8.1) $$\frac{du}{dt} + \nu Au + B(u,u) = f$$

(8.2) $$u(0) = u_0 .$$

The function $f = f(t)$ is a given vector valued function. The function u_0 is given and so is $\nu > 0$. The solution will be a vector valued function $u(t)$ such that $Au(t)$ and $B(u(t),u(t)) = P(u(t)\cdot\nabla u(t))$ make sense. We shall make this more precise later.

The need to study weak solutions arises mainly for d = 3 because even if u_0 and f are very nice functions, in this case the existence of a classical solution of the Navier-Stokes equations is known, in general, only for short time intervals.

The method of proof of existence of solutions to (8.1) that we present is based on Galerkin approximations and energy estimates. First we describe the Galerkin approximations. These are systems of Ordinary Differential Equations. Let m be a positive integer. We consider $w_1,\ldots,w_m$, the first m eigenfunctions of A. (The Galerkin method can be devised starting from a different basis, also.) We consider the projector $P_m : H \to H$ onto the space of $w_1,\ldots,w_m$. Applying P_m to (8.1) would yield the equation

$$\frac{d}{dt} P_m u + \nu A(P_m u) + P_m B(u,u) = P_m f.$$

The Galerkin system of order m is the system

$$(8.3) \qquad \frac{du_m}{dt} + \nu A u_m + P_m B(u_m, u_m) = g_m$$

$$(8.4) \qquad u_m(0) = u_m^0$$

The function $u_m(t)$ belongs to $P_m H$.

More precisely, let us denote by $\xi_j = \xi_j(t)$ the jth component of $u_m(t)$:

$$\xi_j(t) = (u_m(t), w_j).$$

Also, let $\eta_j(t) = (g_m(t), w_j)$ be the components of g_m. Then (8.3) is equivalent to

$$(8.5) \qquad \frac{d\xi_j}{dt} + \nu\lambda_j \xi_j + \sum_{k,\ell=1}^{m} b(w_k, w_\ell, w_j)\xi_k \xi_\ell = \eta_j, \quad j = 1,\ldots,m$$

The initial data u_m^0 has coefficients $(u_m^0, w_j) = \xi_j^0$ and (8.4) is equivalent to

$$(8.6) \qquad \xi_j(0) = \xi_j^0, \quad j = 1,\ldots,m.$$

Thus the Galerkin system of order m is a quadratic, constant coefficient $m \times m$ ODE system. If the function g_m and thus the vector η are time independent the system is autonomous. We shall derive some bounds for solutions of (8.3) and (8.4). Our goal is to have enough control on the solutions of (8.3) to be able to let m tend to infinity and obtain a solution to (8.1). We shall consider an arbitrary $T > 0$ and fix it throughout this section. Let us assume that the function $g_m(t)$ is continuous on $[0,T]$ with values in V'. Since $w_j \in V$ this guarantees that the function $\eta:[0,T] \to \mathbb{R}^m$ is continuous. From the theory of ordinary differential equations we know that (8.5), (8.6) have a unique solution

$\xi(t)$ defined for t in a neighborhood of t = 0. The tensor $b(w_k, w_\ell, w_j)$ is, for each fixed k, antisymmetric in ℓ and j:

$$b(w_k, w_\ell, w_j) = -b(w_k, w_j, w_\ell), \quad 1 \le k, \ell, j \le m.$$

This property implies that the maximal interval of existence of $\xi(t)$ coincides with that of η. In other words, if η would be defined (and continuous) for all t so would be ξ. Indeed, taking the scalar product of (8.5) with ξ in $\mathbb{R}^m$ we obtain

$$(8.7) \qquad \frac{1}{2}\frac{d}{dt}|\xi(t)|^2 + \nu \sum_{j=1}^{m} \lambda_j \xi_j^2(t) = \langle \eta(t), \xi(t) \rangle$$

because

$$(8.8) \qquad \sum_{k,\ell,j} b(w_k, w_\ell, w_j)\xi_k \xi_\ell \xi_j = 0.$$

Together with Gronwall's inequality, (8.7) would prove that $|\xi(t)|$ is bounded as long as $|\eta(t)|$ is bounded. (In particular, for negative t, too.) For positive t we can use the fact that $\lambda_j > 0$ to estimate

$$|\xi|^2 \le |\xi^0|^2 e^{-\nu\lambda_1 t} + \int_0^t e^{-\nu\lambda_1(t-s)} \frac{|\eta(s)|^2}{\nu\lambda_1} ds.$$

Let us estimate the right-hand side of (8.7) differently

$$|\langle \eta(t), \xi(t) \rangle| \le \left(\sum_{j=1}^{m} \lambda_j \xi_j^2\right)^{1/2} \left(\sum_{j=1}^{m} \lambda_j^{-1} \eta_j^2\right)^{1/2}$$

We return to the notation

$$u_m(t) = \sum_{j=1}^{m} \xi_j(t) w_j \; ; \; g_m(t) = \sum_{j=1}^{m} \eta_j(t) w_j .$$

Then we get from (8.7):

(8.9) $$\frac{1}{2}\frac{d}{dt}|u_m|^2 + \nu\| u_m \|^2 \leq |A^{-1/2}g_m|\| v_m \| \leq \frac{\nu}{2}\| v_m \|^2 + \frac{1}{2\nu}|A^{-1/2}g_m|^2$$

Thus we obtain from (8.9) and Gromwall's lemma

(8.10) $$|y_m(t)|^2 \leq |u_m(0)|^2 e^{-\nu\lambda_1 t} + \int_0^t e^{-\nu\lambda_1(t-s)} \frac{|g_m(s)|^2_{V'}}{\nu}$$

We use the notation $|h|_{V'} = |A^{-1/2}h|$, meaning that we identify the dual V' of V with $\mathcal{D}(A^{-1/2})$.

Let us assume that the sequence u_m^0 is bounded in H and that the sequence g_m is bounded in $L^2(0,T;V')$. We denote by $L^p(I,X)$ for $I \subset \mathbb{R}$, $1 \leq p \leq \infty$, X a Banach space, the space of vector valued functions x(t), $x:I \to X$ such that the scalar function $\| x(t) \|_X$ is measurable and $\int_I \|x(t)\|_X^p \, dt < \infty$ for $p < \infty$, $\operatorname{ess\,sup}_{t \in I} \|x(t)\| < \infty$ for $p < \infty$. Thus, if

(8.11) $$\int_0^T |g_m(s)|^2_{V'} ds \leq \int_0^T |g(s)|^2_{V'} ds$$

for some $g \in L^2(0,T,V')$ and if

(8.12) $$|u_m^0| \leq |u^0| \quad \text{for some } u^0$$

as is the case if $g_m = P_m g$, $u_m^0 = P_m u^0$ then (8.10) means that

(8.13) The sequence u_m is bounded in $L^\infty(0,T;H)$.

This fact is not yet sufficient for passing to the limit. Let us integrate (8.9) between 0 and some $t \leq T$.

(8.14) $$|u_m(t)|^2 + \nu\int_0^t \| u_m(s) \|^2 ds \leq |u_m^0|^2 + \frac{1}{\nu}\int_0^t |A^{-1/2}g_m(s)|^2 ds.$$

In particular,

(8.15) $$\nu\int_0^T \| u_m(s) \|^2 ds \leq |u^0|^2 + \frac{1}{\nu}\int_0^T |g(s)|^2_{V'} ds .$$

That is,

(8.16) The sequence u_m is bounded in $L^2(0,T;V)$.

We need to have some uniform bound on du_m/dt. First let us consider the equation (8.3):

$$\frac{du_m}{dt} = -\nu\, Au_m - P_m B(u_m,u_m) + g_m.$$

Since u_m is bounded in $L^2(0,T,V)$ it follows that Au_m is bounded in $L^2(0,T,V')$. Actually,

$$\int_0^T |Au_m(s)|^2_{V'} ds = \int_0^T \| u_m(s) \|^2_{V'} \; ds .$$

Moreover, by assumption g_m is bounded in $L^2(0,T;V')$. It remains to investigate the term $P_m B(u_m,u_m)$. Now

$$|P_m B(u_m,u_m)|_{V'} = |A^{-1/2} P_m B(u_m,u_m)| = |P_m A^{-1/2} B(u_m,u_m)| \le |A^{-1/2} B(u_m,u_m)|.$$

Using (6.20) with $s_3 = 1$, $s_1 = 1/2$, $s_2 = 0$ we get

$$|P_m B(u_m,u_m)|_{V'} \le c|u_m|^{1/2} \| u_m \|^{3/2}.$$

Since $|u_m(s)|$ are bounded uniformly and since $\| u_m(s) \|^2$ are uniformly integrable we obtain

(8.17) $\dfrac{du_m}{dt}$ are bounded in $L^{4/3}(0,T;V')$.

We proved therefore

Lemma 8.1. Let $g_m(t) = \sum_{j=1}^m \eta_j(t) w_j$ and $u_m^0 = \sum_{j=1}^m \xi_j^0 w_j$ be given. Assume that (8.11) and (8.12) hold. Then the solution $u_m(t) = \sum_{j=1}^m \xi_j(t) w_j$ to the Galerkin system (8.3), (8.4) exists and is unique on $[0,T]$.

Moreover, the sequence u_m is uniformly bounded in $L^\infty(0,T;H)$ and in $L^2(0,T;V)$. The sequence du_m/dt is uniformly bounded in $L^{4/3}(0,T;V')$.

All we need now is some kind of compact imbedding theorem, of the type of the Rellich Lemma but for vector valued functions.

Lemma 8.2. Let u_m be a sequence of functions satisfying

$$(8.18) \quad \int_0^T \| u_m(s) \|_V^2 \, ds \leq M$$

$$(8.19) \quad \int_0^T \| \frac{d}{ds} u_m(s) \|_{V'}^p \, ds \leq M$$

for some $0 < M$, $p > 1$ and all $m = 1,2,\ldots$. Then there exists a subsequence u_{m_j} of u_m which is convergent in $L^2(0,T;H)$ to some function u $L^2(0,T;H)$, i.e.,

$$\lim_{j \to \infty} \int_0^T |u_{m_j}(s) - u(s)|_H^2 \, ds = 0.$$

In order to clarify what are the properties needed for this selection theorem we shall prove it in an abstract version. Instead of the spaces $V \subset H \subset V'$ we shall consider three separable reflexive Banach spaces $X_1 \hookrightarrow X_0 \hookrightarrow X_{-1}$.

We will assume that the inclusions $X_1 \hookrightarrow X_0$ is compact and that the inclusion $X_0 \hookrightarrow X_{-1}$ is continous. If $u:[0,T] \to X_0$ is a strongly measurable function we say that du/ds belongs to $L^p(0,T;X_{-1})$ for some $1 < p < \infty$ if there exists $v \in L^p(0,T;X_{-1})$ such that

$$u(t_2) - u(t_1) = \int_{t_1}^{t_2} v(s)ds$$

for almost all t_1, t_2 in $[0,T]$. The element $\int_{t_1}^{t_2} v(s)ds$ can be defined using the duality:

$$L(\int_{t_1}^{t_2} v(s)ds) = \int_{t_1}^{t_2} L(v(s))ds$$

for every $L \in (X_{-1})'$ (the dual of X_{-1}).

Lemma 8.3. For every $\varepsilon > 0$ there exists $c_\varepsilon > 0$ such that, for each $x \in X_1$

$$\|x\|_0 \leq \varepsilon\|x\|_1 + c_\varepsilon\|x\|_{-1}.$$

Proof. Assume not. There exists a sequence $x_m \in X_1$ such that

$$\|x_m\|_0 \geq \varepsilon\|x_m\|_1 + m\|x_m\|_{-1}$$

Take $y_m = x_m/\|x_m\|_1$. It satisfies

$$\|y_m\|_0 \geq \varepsilon + m\|y_m\|_{-1},$$

$\|y_m\|_1 = 1$. Since X_1 is reflexive the unit ball is weakly compact. We may assume (by passing to a subsequence) that y_m converges weakly in X_1 to y. Since the inclusion $X_1 \hookrightarrow X_0$ is compact, y_m converges strongly to y in X_0. Since $\|y_m\|_0$ is bounded it follows that $\|y_m\|_{-1}$ converges to 0. Thus $\|y\|_{-1} = 0$ and y must equal 0. But $\|y_m\|_0 \geq \varepsilon$ implies $\|y\|_0 \geq \varepsilon$, absurd.

Lemma 8.4. Let u_m be a bounded sequence in $L^{p_1}(0,T;X_1)$. Assume that du_m/dt is bounded in $L^{p_2}(0,T;X_{-1})$. Here $1 < p_1 < \infty$, $1 < p_2 < \infty$. Then there exists a subsequence $u_{m'}$ of u_m, converging in $L^{p_1}(0,T;X_0)$.

Proof. The space $L^{p_1}(0,T;X_1)$ is separable, reflexive. (The dual of $L^{p_1}(0,T,X_1) = L^{p_1'}(0,T,X_1')$ where $\frac{1}{p_1} + \frac{1}{p_1'} = 1$, X_1' is the dual of X_1). Therefore there exists a subsequence of u_m which is weakly convergent in $L^{p_1}(0,T;X_1)$. Subtracting the limit we may assume that $u_{m'}$ converges weakly in $L^{p_1}(0,T;X_1)$ to 0. We want to prove that the convergence is strong in $L^{p_1}(0,T;X_0)$. Using Lemma 8.3 we see that it is enough to prove that $u_{m'}$ converges strongly to 0 in $L^{p_2}(0,T;X_{-1})$. Indeed since

$$\|x\|_{X_0}^{p_1} \leq \varepsilon \|x\|_{X_1}^{p_1} + c_{\varepsilon,p_1} \|x\|_{X_{-1}}^{p_1}$$

for all $x \in X_1$, $\varepsilon > 0$ and since

$$\int_0^T \| u_{m'}(t) \|_{X_1}^{p_1} dt$$

is bounded it follows that, for every ε

$$\int_0^T \| u_{m'} \|_{X_0}^{p_1} dt \le \varepsilon \sup_{m'} \int_0^T \| u_{m'}(t) \|_{X_1}^{p_1} dt + c_{\varepsilon,p_1} \int_0^T \| u_{m}'(t) \|_{X_{-1}}^{p_1} dt.$$

Now if I is any subinterval of [0,T] the sequence $\int_I u_m(s)ds$ is weakly convergent in X_1 to 0. Indeed denoting by χ_I the characteristic function of I, for any element L of the dual X_1', $\chi_I(s)L$ is an element of the dual of $L^{p_1}(0,T;X_1)$ and

$$\langle u_m, \chi_I L\rangle = \int_I \langle u_m(s),L\rangle\, ds = \langle \int_I u_m(s)ds, L\rangle$$

converges to zero. Thus since the inclusion $X_1 \hookrightarrow X_0$ is compact we know that $\int_I u_m(s)ds$ converges strongly to zero in X_0 and <u>a fortiori</u> in X_{-1}. Let t be any number in [0,T]. $u_m(t) - u_m(t_1) = \int_{t_1}^t \frac{du_m}{ds} ds$. Let us take an average in t_1 over an interval I_ε of size ε:

$$u_{m'}(t) = \frac{1}{\varepsilon} \int_{t-\varepsilon}^t u_{m'}(t_1)dt_1 + \frac{1}{\varepsilon} \int_{t-\varepsilon}^t (s - t + \varepsilon) \frac{du_{m'}}{ds} ds .$$

Using Hölder's inequality

$$\frac{1}{\varepsilon} \int_{t-\varepsilon}^t |s-t+\varepsilon| \, \| \frac{du_{m'}}{ds} \|_{X_{-1}} ds \le \frac{1}{\varepsilon} \left(\int_{t-\varepsilon}^t (s-t+\varepsilon)^{p_2'} \right)^{1/p_2'} \left(\int_{t-\varepsilon}^t \| \frac{du_{m'}}{ds} \|^{p_2} \right)^{1/p_2}$$

$$\le \left(\frac{1}{p_2' + 1}\right)^{1/p_2'} \varepsilon^{1/p_2'} \left(\int_0^T \| \frac{du_{m'}}{ds} \|^{p_2} \right)^{1/p_2} \le c\varepsilon^{1/p_2'}$$

where $\frac{1}{p_2'} + \frac{1}{p_2} = 1$.

For given $\varepsilon_0 > 0$ choose ε such that $c\varepsilon^{1/p_2'} < \varepsilon_0/2$. Then fix ε; it follows that

$$\| u_{m'}(t) \|_{X_{-1}} \le \frac{\varepsilon_0}{2} + \frac{1}{\varepsilon} \| \int_{t-\varepsilon}^t u_{m'}(t_1)dt_1 \|_{X_{-1}}$$

Since the second term on the right-hand side of this inequality tends to zero as $m' \to \infty$ we proved that, for each t, $\|u_{m'}(t)\|_{X_{-1}} \xrightarrow{m'} 0$. The result follows from the Lebesgue dominated convergence theorem.

Remark. Actually we proved also that $u_{m'}$ converges to zero in $C[0,T;X_{-1}]$ i.e., that $\sup_{t \in [0,T]} \|u_{m'}(t)\|_{X_{-1}} \to 0$. This follows from the fact that there is pointwise convergence and the uniform Hölder continuity

$$\|u_m(t_0) - u_m(t_1)\|_{X_{-1}} \le c|t_0 - t_1|^{1/p_2'}$$

Definition 8.5. A weak solution of the Navier-Stokes equations (8.1), (8.2) is a function $u \in L^2(0,T;V) \cap C_w(0,T;H)$ satisfying $\frac{du}{dt} \in L^1_{loc}(0,T,V')$ and

(8.20) $$\langle \frac{du}{dt}, v\rangle + \nu((u,v)) + b(u,u,v) = (f,v) \text{ a.e. in } t, \text{ for all } v \in V.$$

(8.21) $$u(0) = u_0$$

The space $C_w(0,T;H)$ is a subspace of $L^\infty(0,T;H)$ consisting of functions which are weakly continuous: $(u(t),h)$ is a continuous function, for all $h \in H$. In particular (8.21) is taken in this sense.

Theorem (Leray). There exists at least a weak solution of (8.1), (8.2) for every $u_0 \in H$, $f \in L^2(0,T;V')$. Moreover,

$$\frac{du}{dt} \in L^{4/3}(0,T;V') \quad \text{for } d = 3$$

$$\frac{du}{dt} \in L^2(0,T;V') \quad \text{for } d = 2$$

and the energy inequality

(8.22) $$\frac{1}{2}|u(t)|^2 + \nu\int_{t_0}^{t} \|u(s)\|^2 ds \le \frac{1}{2}|u(t_0)|^2 + \int_{t_0}^{t} < f(s),u(s)> ds$$

holds for all $0 \le t_0 \le t \le T$, t_0 a.e. in $[0,T]$.

Proof. Let $u_m(t)$ be solutions of the Galerkin equations (8.3), (8.4) with $g_m = P_m f$, $u_m^0 = P_m u^0$. From Lemma 8.1 and 8.4 we may assume that $u_{m'}$, a subsequence of u_m converges weakly in $L^2(0,T;V)$ strongly in $L^2(0,T,H)$ and in $C[0,T;V']$ to some u. Moreover, $du_{m'}/dt$ converges weakly in $L^{4/3}(0,T;V')$ to du/dt. Let $v \in V$ be arbitrary. Take the scalar product of (8.3) with v and integrate

$$(u_{m'}(t),v) + \nu\int_{t_0}^{t}((u_{m'}(s),v))ds + \int_{t_0}^{t} b(u_{m'}(s),u_{m'}(s),P_m v)ds$$
$$= (u_{m'}(t_0),v) + \int_{t_0}^{t} <f(s),P_m v>ds.$$

Since $u_{m'}$ converges to u in $L^2(0,T;V)$ weakly, by extracting a subsequence, relabeled u_m, we may assume that $u_m(t_0)$ converges to $u(t_0)$ weakly in V for all $t_0 \in [0,T] \setminus E$, for some E of Lebesgue measure 0. Thus $\lim_{m\to\infty} u_m(t_0) = u(t_0)$ strongly in H, for $t_0 \notin E$. Now clearly

$$\lim_{m\to\infty} \int_{t_0}^{t}((u_m(s),v))ds = \int_{t_0}^{t}((u(s),v))ds.$$

A simple argument shows that

$$\lim_{m\to\infty}\int_{t_0}^{t} b(u_m(s),u_m(s),P_m v)ds = \int_{t_0}^{t} b(u(s),u(s),v)ds.$$

It follows that for $t \ge t_0$ and $t,t_0 \notin E$

$$(u(t),v)-(u(t_0),v)+ \nu\int_{t_0}^{t}((u(s),v))ds+\int_{t_0}^{t} b(u(s),u(s),v)ds = \int_{t_0}^{t} <f(s),v>ds.$$

This implies the weak continuity of $u(t)$ in H because V is dense in H and $\sup_{t\in[0,T]} |u(t)|$ is finite. Also the last relation implies (8.20).

For the energy inequality take the inequalilty

$$\frac{1}{2}|u_m(t)|^2 + \nu\int_{t_0}^{t} \|u_m(s)\|^2 \leq \frac{1}{2}|u_m(t_0)|^2 + \int_{t_0}^{t} \langle f(s),u_m(s)\rangle\, ds.$$

Assume that $t_0 \notin E$. The right-hand side has for limit as $m \to \infty$,

$\frac{1}{2}|u(t_0)|^2 + \int_{t_0}^{t} \langle f(s),u(s)\rangle ds$. Passing to $\overline{\lim}_{m\to\infty}$ using

$\overline{\lim}\,(a_m + b_m) \geq \overline{\lim}\, a_m + \underline{\lim}\, b_m$ and the fact that if $x_m \to x$ weakly in a Hilbert space X then $\|x\| \leq \underline{\lim}\,\|x_m\|$ we obtain the energy inequality

$$\frac{1}{2}|u(t)|^2 + \nu\int_{t_0}^{t} \|u(s)\|^2 ds \leq \frac{1}{2}|u(t_0)|^2 + \int_{t_0}^{t} \langle f(s),u(s)\rangle ds$$

for $t_0 \notin E$, $t \geq t_0$.

The fact that du/dt belongs to $L^2(0,T,V')$ for d = 2 follows from the fact that one can estimate in d = 2 $|A^{-1/2}B(u,u)| \leq c|u|\|u\|$. Indeed $(A^{-1/2}B(u,u),v) = B(u,u,A^{-1/2}v) = -\, b(u,A^{-1/2}v,u)$ and thus $|(A^{-1/2}B(u,u),v)| \leq c|u|\|u\||v|$. This together with the fact that u_m is bounded in $L^2(0,T,V)$ and $L^\infty(0,T,H)$ makes $A^{-1/2}B(u_m,u_m)$ bounded in $L^2(0,T,V')$.

9

STRONG SOLUTIONS

Let u_m be a solution of the Galerkin system

(9.1) $$\frac{du_m}{dt} + \nu A u_m + P_m B(u_m, u_m) = P_m f$$

(9.2) $$u_m(0) = P_m u_0$$

We take the scalar product of (9.1) with u_m and obtain (see (8.8))

$$\frac{1}{2}\frac{d}{dt}|u_m|^2 + \nu\|u_m\|^2 = (f, u_m) \leq \frac{|f|^2}{2\nu\lambda_1} + \frac{\nu\lambda_1}{2}|u_m|^2$$

since $\lambda_1 |u_m|^2 \leq \|u_m\|^2$ it follows that

(9.3) $$\frac{d}{dt}|u_m|^2 + \nu\|u_m\|^2 \leq \frac{|f|^2}{\nu\lambda_1}$$

and thus

(9.4) $$\nu \int_0^t \|u_m\|^2 ds \leq |u_0|^2 + \int_0^t \frac{|f|^2}{\nu\lambda_1} ds$$

and

(9.5) $$|u_m(t)|^2 \leq |u_0|^2 e^{-\nu\lambda_1 t} + \int_0^t e^{-\nu\lambda_1(t-s)} \frac{|f|^2}{\nu\lambda_1} ds.$$

Let us take the scalar product of (9.1) with Au_m. We get

(9.6) $$\frac{1}{2}\frac{d}{dt}\|u_m\|^2 + \nu|Au_m|^2 + b(u_m, u_m, Au_m) = (f, Au_m).$$

We will describe separately the cases of spatial dimensions two and three.

The two dimensional case

Our aim is to give a bound on $\sup_{0<t} \|u_m(t)\|$ which is independent on m. For this purpose let us first assume that $f \in L^\infty(\mathbb{R}_+,H)$ and denote by

(9.7) $$|f|_\infty = \sup_{t>0} |f(t)|.$$

The estimates (9.4) and (9.5) become

(9.8) $$\nu \int_0^t \|u_m(s)\|^2 \leq |u_0|^2 + \frac{|f|_\infty^2}{\nu\lambda_1} t$$

(9.9) $$|u_m(t)|^2 \leq |u_0|^2 + \frac{|f|_\infty^2}{\nu^2\lambda_1^2}.$$

Integrating (9.3) between t and t + τ and using (9.9) we get

(9.10) $$\nu \int_t^{t+\tau} \|u_m(s)\|^2 ds \leq |u_0|^2 + \frac{|f|_\infty^2}{\nu\lambda_1}(\tau + \frac{1}{\nu\lambda_1}).$$

Therefore, denoting by λ the Lebesgue measure on $\mathbb{R}$, we get

$$\lambda(\{s \ [t,t+\tau] \mid \|u_m(s)\| \geq \rho\} \leq (\frac{|u_0|^2}{\nu} + \frac{|f|_\infty^2}{\nu^2\lambda_1}(\tau + \frac{1}{\nu\lambda_1}))\rho^{-2}$$

Let us take $\rho = \sqrt{2}[\frac{|u_0|^2}{\nu} + \frac{|f|_\infty^2}{\nu^2\lambda_1}(\tau + \frac{1}{\nu\lambda_1})]^{1/2}/\sqrt{\tau}$. It follows that

$$\lambda\{s|s \in [t,t+\tau], \|u_m(s)\| \geq \rho\} \leq \frac{1}{2}\tau$$

and therefore that

In every interval of length τ there exists a time $t_0 \in [t,t+\tau]$ such that

(9.11) $$\|u_m(t_0)\|^2 \leq \frac{2}{\tau}[\frac{|u_0|^2}{\nu} + \frac{|f|_\infty^2}{\nu^2\lambda_1}(\tau + \frac{1}{\nu\lambda_1})].$$

Remark that until now the reasoning did not depend on the spatial dimension d. The two sides of (9.11) scale like L^d/T^2. Now let us estimate in (9.6) the term $|(f,Au_m)| \le \frac{\nu}{4}|Au_m|^2 + \frac{|f|_\infty^2}{\nu}$ and the term $b(u_m,u_m,Au_m)$ using (6.9) with $s_1 = 1/2$, $s_2 = 1/2$, $s_3 = 0$.

$$|b(u_m,u_m,Au_m)| \le c|u_m|^{1/2}\|u_m\|\,|Au_m| \tag{9.12}$$

Thus (9.6) becomes

$$\frac{1}{2}\frac{d}{dt}\|u_m\|^2 + \nu|Au_m|^2 \le \frac{\nu}{4}|Au_m|^2 + \frac{|f|_\infty^2}{\nu} + \frac{\nu}{4}|Au_m|^2 + \frac{c}{\nu^3}|u_m|^2\|u_m\|^4$$

$$\frac{d}{dt}\|u_m\|^2 + \nu|Au_m|^2 \le 2\frac{|f|_\infty^2}{\nu} + \frac{c}{\nu^3}|u_m|^2\|u_m\|^4. \tag{9.13}$$

Multiplying with $\exp -\int_{t_0}^{t}\frac{c}{\nu^3}|u_m|^2\|u_m\|^2 ds$ we obtain

$$\frac{d}{dt}\left[\|u_m\|^2\exp -\int_{\tau_0}^{t}\frac{c}{\nu^3}|u_m|^2\|u_m\|^2 ds\right] \le \frac{2|f|_\infty^2}{\nu}\exp -\int_{t_0}^{t}\frac{c}{\nu^3}|u_m|^2\|u_m\|^2 ds$$

Integrating we obtain

$$\|u_m\|^2 \le \|u_m(t_0)\|^2\exp\left(\int_{t_0}^{t}\frac{c}{\nu^3}|u_m|^2\|u_m\|^2 ds\right) + \frac{2|f|_\infty^2}{\nu}(t-t_0)\exp\int_{t_0}^{t}\frac{c}{\nu^3}|u_m|^2\|u_m\|^2 ds \tag{9.14}$$

The inequality (9.14) is valid for $0 \le t_0 \le t$. Now we estimate the exponential using (9.10) and (9.9):

$$\frac{c}{\nu^3}\int_{t_0}^{t}|u_m|^2\|u_m\|^2 ds \le \frac{c}{\nu^4}\left[|u_0|^2 + \frac{|f|_\infty^2}{\nu^2\lambda_1^2}\right]\left[|u_0|^2 + \frac{|f|_\infty^2}{\nu\lambda_1}\left(t - t_0 + \frac{1}{\nu\lambda_1}\right)\right]$$

If t_0 is chosen in the interval $[t-\tau,t]$ (assuming $t \ge \tau$) such that (9.11) be valid we get

$$\|u_m(t)\|^2 \le Ae^B + \frac{2|f|_\infty^2}{\nu}\tau e^B$$

where A is the right-hand side of (9.11) and B is

$$\frac{c}{\nu^4}[|u_0|^2 + \frac{|f|_\infty^2}{\nu^2\lambda_1^2}][|u_0|^2 + \frac{|f|_\infty^2}{\nu\lambda_1}(\tau + \frac{1}{\nu\lambda_1})].$$

Assume now that $\tau \le 1/\nu\lambda_1$. Then we get, for $t \ge \tau$

$$(9.15)\qquad \|u_m(t)\|^2 \le (\frac{2}{\tau}[\frac{|u_0|^2}{\nu} + \frac{2|f|_\infty^2}{\nu^3\lambda_1^2}] + \frac{2|f|_\infty^2}{\nu^2\lambda_1})\exp[\frac{c}{\nu^4}(|u_0|^2 + \frac{|f|_\infty^2}{\nu^2\lambda_1^2})^2]$$

This bound is not only uniform for $m \in N$ and $t \ge \tau$ but also does not involve the quantity $\|u_0\|$. In other words, although $\|u_m(0)\|$ are finite, $\sup_m \|u_m(0)\|$ is not required to be finite. If we allow $\|u_0\|$ to enter the estimate we can fix $\tau = 1/\nu\lambda_1$ for instance and use (9.14) with $t_0 = 0$ to estimate $\|u_m(t)\|$ for $0 \le t \le 1/\nu\lambda_1$. However, we can avoid this in the following way: Let us use (9.15) on an interval $[\frac{1}{\nu\lambda_1}\cdot\frac{1}{2^{k+1}}, \frac{1}{\nu\lambda_1}\cdot\frac{1}{2^k}]$ with $\tau = \frac{1}{\nu\lambda_1}\cdot\frac{1}{2^{k+1}}$. If $t \le \frac{1}{\nu\lambda_1}\cdot\frac{1}{2^k}$ then $\tau \le t \le 2\tau$ and thus (9.15) implies, for $t \in [\frac{1}{2^{k+1}}\frac{1}{\nu\lambda_1}, \frac{1}{2^k}\frac{1}{\nu\lambda_1}]$ that

$$(9.16)\qquad t\|u_m(t)\|^2 \le (4[\frac{|u_0|^2}{\nu} + \frac{2|f|_\infty^2}{\nu^3\lambda_1^2}] + \frac{2|f|_\infty^2}{\nu^3\lambda_1^2})\exp[\frac{c}{\nu^4}(|u_0|^2 + \frac{|f|_\infty^2}{\nu^2\lambda_1^2})^2].$$

Since the estimate (9.16) is independent of k we proved that (9.16) is true for any t, $0 < t \le 1/\nu\lambda_1$. Summarizing we obtain:

<u>Proposition 9.1.</u> Let $m \ge 1$ be an integer. Let $u_0 \in H$, $f \in L^\infty(\mathbb{R},H)$. Let u_m be the solution of the Galerkin system

$$\begin{cases} \frac{du_m}{dt} + \nu A u_m + P_m B(u_m,u_m) = P_m f \\ u_m(0) = P_m u_0 \end{cases}$$

There exists a constant ρ_0 depending on ν, λ_1, $|u_0|$ and $\sup_{t\ge 0}|f(t)|$ only such that

(9.17) $$\sup_{0<t\leq 1/\nu\lambda_1} \nu\lambda_1 t\|u_m(t)\|^2 \leq \rho_0^2$$

(9.18) $$\sup_{t\geq 1/\nu\lambda_1} \|u_m(t)\| \leq \rho_0.$$

If $u_0 \in V$, there exists a constant ρ_1 depending on $\|u_0\|$, ν, λ_1 and $\sup_t |f(t)|$ only such that, in addition to (9.18),

(9.19) $$\sup_{0\leq t\leq 1/\nu\lambda_1} \|u_m(t)\| \leq \rho_1$$

holds.

Passing to the limit we obtain

Theorem 9.2. Let Ω be an open bounded set of class C^2 included in $\mathbb{R}^2$. Let $u_0 \in H$, $f \in L^\infty(\mathbb{R}_+, H)$. Let $T > 0$. There exists a solution u of the Navier-Stokes equation

$$\begin{cases} \dfrac{du}{dt} + \nu Au + B(u,u) = f \\ u(0) = u_0 \end{cases}$$

satisfying $u \in L^\infty_{loc}(0,T;V) \cap L^2_{loc}(0,T;\mathcal{D}(A)) \cap L^\infty(0,T,H) \cap L^2(0,T;V)$. Moreover,

(9.20) $$\sup_{0<t\leq 1/\nu\lambda_1\leq T} \nu\lambda_1 t\|u(t)\|^2 + \sup_{1/\nu\lambda_1\leq t\leq T} \|u(t)\|^2 \leq 2\rho_0^2$$

where ρ_0 depends on $|u_0|, \nu, \lambda_1$, and $\sup_t |f(t)|$ but not on T.

If u_0 belongs to V then $u \in L^\infty(0,T;V) \cap L^2(0,T, \mathcal{D}(A))$ and

$$\sup_{0\leq t\leq T} \|u(t)\|^2 + \nu \int_0^T |Au|^2 dt \leq C\left(\frac{1}{\nu\lambda_1} + T\right)$$

with C depending on ν, λ_1, $\|u_0\|$ and $\sup_t |f(t)|$ but not on T.

An upper bound for ρ_0 is

$$\rho_0^2 \leq 10(\lambda_1|u_0|^2 + \frac{|f|_\infty^2}{\nu^2\lambda_1})\exp[\frac{c}{\nu^4}(|u_0|^2 + \frac{|f|_\infty^2}{\nu^2\lambda_1^2})^2]. \tag{9.21}$$

Let us consider now

The three dimensional case.

We return to the Galerkin system (9.1), (9.2). Let us assume that $f \in L^2(0,T;H)$ for some $T > 0$ and $u_0 \in V$. We shall estimate in (9.6) $|(f,Au_m)| \leq \frac{\nu}{4}|Au_m|^2 + \frac{|f|^2}{\nu}$. In order to estimate the term $b(u_m,u_m,Au_m)$ we use (6.9) with $s_1 = 1$, $s_2 = 1/2$, $s_3 = 0$

$$|b(u_m,u_m,Au_m)| \leq c\|u_m\|^{3/2}|Au_m|^{3/2} \tag{9.22}$$

It follows from (9.6) that

$$\frac{1}{2}\frac{d}{dt}\|u_m\|^2 + \nu|Au_m|^2 \leq \frac{\nu}{4}|Au_m|^2 + \frac{|f|^2}{\nu} + c\|u_m\|^{3/2}|Au_m|^{3/2}$$

$$\leq \frac{\nu}{2}|Au_m|^2 + \frac{|f|^2}{\nu} + \frac{c}{\nu^3}\|u_m\|^6$$

Thus

$$\frac{d}{dt}\|u_m\|^2 + \nu|Au_m|^2 \leq \frac{2|f|^2}{\nu} + \frac{c}{\nu^3}\|u_m\|^6 \tag{9.23}$$

Let us assume that

$$\|u_m(0)\|^2 + \frac{2}{\nu}\int_0^T |f(t)|^2 dt \leq \frac{c^{-1/2}}{2}\nu^2\lambda_1^{1/2} \tag{9.24}$$

Then we claim, for all $T \geq t \geq 0$

$$\|u_m\|^2 < c^{-1/2}\nu^2\lambda_1^{1/2}. \tag{9.25}$$

Indeed, from (9.24) it follows that $\|u_m(0)\|^2 < c^{-1/2}\nu^2\lambda_1^{1/2}$. Since $\|u_m(t)\|$ is smooth $\|u_m(t)\|^2 < c^{-1/2}\nu^2\lambda_1^{1/2}$ for small t. As long as $\|u_m(t)\|^2 < c^{-1/2}\nu^2\lambda_1^{1/2}$ it follows that $\nu|Au_m|^2 - \frac{c}{\nu^3}\|u_m\|^6$ is positive.

Indeed,

$$\nu|Au_m|^2 - \frac{c}{\nu^3}\|u_m\|^6 \geq \nu\lambda_1\|u_m\|^2 - \frac{c}{\nu^3}\|u_m\|^6$$

$$= \nu\lambda_1\|u_m\|^2[1 - \frac{c}{\nu^4\lambda_1}\|u_m\|^4] > 0.$$

From (9.23) it follows that, as long as $\|u_m(t)\|^2 < c^{-1/2}\nu^2\lambda_1^{1/2}$

$$\|u_m(t)\|^2 \leq \frac{2}{\nu}\int_0^t |f|^2 + \|u_m(0)\|^2 \leq \frac{2}{\nu}\int_0^T |f|^2 + \|u_m(0)\|^2 \leq \frac{c^{-1/2}}{2}\nu^2\lambda_1^{1/2}.$$

Therefore the least upper bound of the set of $t \leq T$ such that (9.25) is satisfied must be T.

Passing to the limit in m we obtain:

Theorem 9.3. Let Ω be an open bounded set in $\mathbb{R}^3$ of class C^2. There exists a scale independent positive constant C such that, for $u_0 \in V$ and $f \in L^2(0,T;H)$ satisfying

$$(9.26) \qquad \frac{\|u_0\|^2}{\nu^2\lambda_1^{1/2}} + \frac{2}{\nu^3\lambda_1^{1/2}}\int_0^T |f(t)|^2 dt \leq \frac{1}{4\sqrt{C}}$$

there exists a solution u(t) of

$$(9.27) \qquad \frac{du}{dt} + \nu Au + B(u,u) = f$$

$$(9.28) \qquad u(0) = u_0$$

belonging to $L^\infty(0,T;V) \cap L^2(0,T, \mathcal{D}(A))$ and satisfying

$$(9.29) \qquad \frac{\|u(t)\|^2}{\nu^2\lambda_1^{1/2}} + \frac{1}{\nu\lambda_1^{1/2}}\int_0^T |Au(s)|^2 ds \leq \frac{2}{\sqrt{C}}$$

for all $0 \leq t \leq T$.

The condition (9.26) is a nondimensional smallness condition. It can be interpreted in various ways: small initial data and f but arbitrary T, or large ν but arbitrary data and T. It would appear from it that $\|u_0\|^2$ small with respect to $\nu^2\lambda_1^{1/2}$ is necessary for local existence. This, of course, is not the case. Indeed, in the inequality (9.23) we have, ignoring $\nu|Au_m|^2$

$$\frac{d}{dt}\|u_m\|^2 \leq \frac{2|f|^2}{\nu} + \frac{c}{\nu^3}\|u_m\|^6. \tag{9.30}$$

Let us introduce the nondimensional quantities

$$\tilde{y}(t) = \frac{\|u_m(t)\|^2}{\nu^2\lambda_1^{1/2}}$$

$$\tilde{g}(t) = \frac{|f|^2(t)}{\nu^4\lambda_1^{3/2}}$$

Then from (9.30) we get

$$\frac{1}{\nu\lambda_1}\frac{d\tilde{y}}{dt} \leq 2\tilde{g}(t) + c\tilde{y}^3(t)$$

Introducing the nondimensional time $s = \nu\lambda_1 t$ we get, for $y(s) = \tilde{y}(t) = \tilde{y}(\frac{s}{\nu\lambda_1})$, $g(s) = \tilde{g}(\frac{s}{\nu\lambda_1})$

$$\frac{dy}{ds} \leq 2g(s) + cy^3(s) \tag{9.31}$$

If $y(s_0)$ is finite then for s near s_0, y(s) will be finite. Indeed, dividing by $(1 + y)^3$ we get, after integration $\int_{s_0}^{s} d\sigma$

$$\frac{1}{(1+y(s_0))^2} - \frac{1}{(1+y(s))^2} \leq 4\int_{s_0}^{s} g(\sigma)d\sigma + 2c(s - s_0) = E(s,s_0)$$

and therefore

$$(1 + y(s)) \leq \frac{1 + y(s_0)}{\sqrt{1 - (1 + y(s_0))^2 E(s,s_0)}} .$$

Then

$$y(s) \leq \frac{1 + y(s_0) - \quad 1 - (1 + y(s_0))^2 E(s,s_0)}{1 - (1 + y(s_0))^2 E(s,s_0)} .$$

We need to impose the restriction $(1 + y(s_0))^2 E(s,s_0) < 1$. Let us assume thus

$$4 \int_{s_0}^{s} g(\sigma)d\sigma + 2c(s - s_0) \leq \frac{1}{2(1 + y(s_0))^2}$$

Then $y(s) \leq \sqrt{2}\,(y(s_0) + 1)$.

We proved:

Theorem 9.4. Let $\Omega \subset \mathbb{R}^3$ be a bounded open set of class C^2. Let $u_0 \in V$. Let $f \in L^2(0,T_0;H)$ satisfy

$$(9.32) \qquad \int_0^{T_0} \frac{|f|^2}{\nu^4 \lambda_1^{3/2}} \nu\lambda_1 \, dt + \nu\lambda_1 T_0 \leq \frac{1}{4} \; \frac{(4 + 2c)^{-1}}{\left[1 + \dfrac{\|u_0\|^2}{\nu^2 \lambda_1^{1/2}}\right]^2} .$$

Then there exists a solution $u(t)$ of (9.27), (9.28) belonging to $L^\infty(0,T_0;V) \cap L^2(0,T_0; \mathcal{D}(A))$ satisfying

$$(9.33) \qquad \frac{\|u(t)\|^2}{\nu^2 \lambda_1^{1/2}} \leq \sqrt{2}\,\left(\frac{\|u_0\|^2}{\nu^2 \lambda_1^{1/2}} + 1\right)$$

for $0 \leq t \leq T_0$.

10

FURTHER RESULTS CONCERNING WEAK AND STRONG SOLUTIONS

We address first the question of uniqueness of weak solutions. In two dimensions weak solutions are unique.

Theorem 10.1. Let $\Omega \subset \mathbb{R}^2$ be open, bounded, of class C^2. Let $f \in L^2(0,T,V')$. Two solutions belonging to $L^2(0,T;V) \cap C_w(0,T;H)$ of

$$\frac{du}{dt} + \nu Au + B(u,u) = f \tag{10.1}$$

$$u(0) = u_0 \in H \tag{10.2}$$

must coincide.

Proof. Let us call the two solutions u_j, $j = 1,2$. Let w denote their difference $w = u_1 - u_2$. We obtain for w the equation

$$\frac{dw}{dt} + \nu Aw + B(u_1,w) + B(w,u_2) = 0 \tag{10.3}$$

$$w(0) = 0 \tag{10.4}$$

Taking the scalar product of (10.3) with w we obtain

$$\langle \frac{dw}{dt}, w\rangle + \nu\|w\|^2 + B(w,u_2,w) = 0 \tag{10.5}$$

Indeed, from (10.3) and the assumptions on f, u_1, u_2 it follows that $\frac{du_j}{dt}$, j = 1,2 and thus dw/dt all belong to $L^2(0,T,V')$. Then (10.5) takes place almost for every t. Moreover, because of the estimate

$$|B(w,u_2,w)| \leq c|w|\|w\|\,\|u_2\|$$

we see that $\langle\frac{dw}{dt},w\rangle = \frac{1}{2}\frac{d}{dt}|w|^2$ belongs to $L^1(0,T)$. From (10.5) we deduce

$$\frac{d}{dt}|w|^2 \leq \frac{c}{\nu}\|u_2\|^2|w|^2$$

and by Gronwall's inequality

$$(10.6) \qquad |w(t)|^2 \leq |w(0)|^2 \exp \frac{c}{\nu}\int_0^t \|u_2\|^2 ds.$$

Since $|w(0)| = 0$ it follows $w \equiv 0$. The theorem is proven.

This theorem together with the existence theorems settle the situation in the case of two-dimensional Navier-Stokes in a satisfactory manner. If u_0 is in H there exists a unique global weak solution. At time 0+ (i.e., for any t > 0) this solution becomes strong provided the forcing term is in H. Genuine weak solutions are generated only by singular forcing terms.

In the three-dimensional case, the uniqueness of the weak solutions is not known. What is known is the uniqueness of strong solutions.

Theorem 10.2. Let $\Omega \subset \mathbb{R}^3$ be open, bounded, of class C^2. Let $f \in L^2(0,T;H)$ and $u_0 \in V$. Two solutions belonging to $L^2(0,T;\mathcal{D}(A)) \cap C_w(0,T;V)$ of (10.1), (10.2) must coincide.

Proof. One proceeds as in the proof of Theorem (10.1). The term $B(w,u_2,w)$ in (10.5) can be estimated by ((6.9) with $s_1 = s_2 = s_3 = 1/2$)

$$|B(w,u_2,w)| \leq c|w|\ \|w\|\ \|u\|^{1/2}|Au|^{1/2}$$

The equation (10.5) yields

$$\text{(10.7)} \qquad \frac{d}{dt}|w|^2 \leq \frac{c}{\nu}\|u_2\|\ |Au_2|\,|w|^2.$$

From Gronwall's inequality

$$|w(t)|^2 \leq |w(0)|^2 \exp \int_0^t \frac{c}{\nu}\|u_2(s)\|\ |Au_2(s)|ds$$

Since $u_2 \in L^\infty(0,T;V) \cap L^2(0,T;\mathcal{D}(A))$ the integral is finite. The proof is concluded by noting that $w(0) = 0$ implies $w(t) = 0$ for $0 \leq t \leq T$.

Remark 10.3. a) It is clear from this formal proof that we can assume that only one of the two solutions is strong, the other being only a weak solution: in other words, strong solutions are unique in the larger class of weak solutions.

b) The method of proof gives sufficient conditions for uniqueness. For instance, estimating the term $B(w,u_2,w)$ differently

$$|B(w,u_2,w)| \leq c|w|^{1/2}\|w\|^{3/2}\|u_2\ \| \leq \frac{\nu}{2}\|w\|^2 + \frac{c}{\nu^3}|w|^2\|u_2\|^4$$

we obtain $\frac{d}{dt}|w|^2 \leq \frac{c}{\nu^3}\|u_2\|^4|w|^2$ and therefore, uniqueness provided $\int_0^T \|u_2\|^4 dt < \infty$, i.e., $u_2 \in L^4(0,T;V)$

The rest of this sections will be devoted to a few of the aspects of the basic question: do solutions of three-dimensional Navier-Stokes equations lose regularity or not? Suppose u_0 is a very nice function, say $u_0 \in V$. Suppose $\Omega \subset \mathbb{R}^3$ is a bounded set with smooth boundary. Moreover, suppose that the driving force f is time independent and very smooth, (e.g. f = 0!). Then we know, for fixed $\nu > 0$ that there exists a solution $u(t)$ for $0 \leq t \leq T_0$ of (10.1), (10.2) which is a strong solution, i.e., $u_0 \in C_w(0,T_0;V) \cap L^2(0,T;\mathcal{D}(A))$. In particular $\sup_{0 \le s \le T_0} \|u(s)\|$ is finite.

Let $T > T_0$. We know that, starting from u_0 there exists a weak solution $\tilde{u} \in L^2(0,T;V) \cap C_w(0,T;H)$. Moreover, from the preceding Remark 10.3 a) it follows that $\tilde{u}$ and u coincide on $[0,T_0]$. Let us consider for fixed u_0, f and ν the maximal time of existence of a strong solution $T_* = \text{Max}\{T > 0$; there exist $u \in L^\infty(0,T;V) \cap L^2(0,T;\mathcal{D}(A)$, solution of (10.2), (10.2)$\}$. Because of the uniqueness of strong solutions if $T_* < \infty$ then $\limsup_{t \to T_*} \|u(t)\| = \infty$ and $\sup_{t \le T_0} \|u(t)\| < \infty$ for all $T_0 < T_*$. Indeed, if $\limsup_{t \to T_*} \|u(t)\| < \infty$ then for an appropriate constant c and t as close to T_* as we please we would have $\|u(t)\| \le c$. From the local existence theorem (Theorem 9.4) we would be able to find v(s) solution (10.1) and $v(0) = u(t_0) \in V$. The solution v(s) of (10.1) would be a strong solution defined for a time interval $[0,T_0]$ whose length depends on the size of $|f|$ and of $\|v(0)\| = \|u(t_0)\|$ (see (9.32)) Since $\|u(t_0)\|$ is bounded from above as $t_0 \to T_*$ the corresponding T_0 is bounded below, i.e., can be chosen uniformly for t_0 near T_*. If $T_* - t_0 < T_0$, we obtain, since f is time independent, a strong solution $\tilde{u}(s) = v(s + t_0)$ of (10.1) which in view of Theorem 10.2 coincides with u for $t < T_*$. We were able thus to extend u(t) beyond T_* (namely to $T_0 + t_0 > T_*$) contradicting the definition of T_*.

We see thus that the quantity $\|u(\cdot)\|$ becoming infinite is a necessary condition for loss of regularity. One could ask oneself: why identify "regularity" with "strong solutions"? After all, a strong solution is not yet a classical solution of the Navier-Stokes system. The reason is that if a solution is strong then u(t) will be as smooth as u_0 provided the boundary $\partial\Omega$ and f are smooth enough.

We will treat this question in absence of boundaries. In this case, as we pointed out in Remark 4.13, the operator A coincides on its domain with $(-\Delta)P = P(-\Delta)$.

Lemma 10.4. Let u,v be two finite sums

$$u = \sum_{k\in Z^n} u_k \exp\left(\frac{2\pi i}{L}\langle x,k\rangle\right)$$

$$v = \sum_{k\in Z^n} v_k \exp\left(\frac{2\pi i}{L}\langle x,k\rangle\right)$$

satisfying $u_0 = v_0 = 0$, $\bar{u}_k = u_{-k}$, $\bar{v}_k = v_{-k}$ $(u_k, v_k \in C^n)$ and $\langle u_k,k\rangle = 0 = \langle v_k,k\rangle$. Let $s > \frac{n}{2}$ be a real number. Then, denoting $B(u,v) = P(u\cdot\nabla v)$, we have the estimates

$$(10.8) \qquad |A^{s/2}B(u,v)| \le cL^{s-\frac{n}{2}}|A^{s/2}u||A^{\frac{s+1}{2}}v|, \qquad \text{for } s > \frac{n}{2}$$

$$(10.9) \qquad |(A^sB(u,v),v)| \le cL^{s-1-\frac{n}{2}}|A^{s/2}u||A^{s/2}v|^2, \quad \text{if } s > \frac{n}{2}+1.$$

Proof. Without loss of generality we may assume L = 1 since (10.8) and (10.9) are scale invariant. Since A and P commute we can estimate the left hand side of (10.8) as $\sup_{\substack{w\in H\\|w|=1}} (A^{s/2}B(u,v),w)$

$$A^{s/2}B(u,v) = P(-\Delta)^{s/2}(u\cdot\nabla)v$$

since Pw = w we have

$$(A^{s/2}B(u,v),w) = ((-\Delta)^{s/2}(u\cdot\nabla v),w)$$

Now

$$(u\cdot\nabla)v = \sum_{k\in Z^n}(2\pi i)\Big(\sum_{j+\ell=k}\langle u_j,\ell\rangle v_\ell\Big)e^{2\pi i\langle x,k\rangle}$$

so

$$(A^{s/2}B(u,v),w) = \sum_{\substack{j+\ell+k=0 \\ j,\ell,k \in \mathbb{Z}^n}} \langle u_j,\ell\rangle \langle v_\ell, w_k\rangle (4\pi^2|k|^2)^{s/2}(2\pi i)$$

Now we use the inequality

$$(10.10) \qquad |j+\ell|^\alpha \leq (|j|+|\ell|)^\alpha \leq C_\alpha[|j|^\alpha + |\ell|^\alpha]$$

valid for any $j,\ell \in \mathbb{R}^n$, $\alpha > 0$. The first part is trivial and the second is just the boundedness of $f(x) = \dfrac{(x+1)^\alpha}{1+x^\alpha}$, $x \geq 0$. Therefore we estimate

$$|(A^{s/2}B(u,v),w)| \leq (2\pi)^{s+1}c_s \sum_{j+\ell+k=0} |u_j||v_\ell||w_k||\ell|[|\ell|^s + |j|^s]$$

$$= (2\pi)^{s+1}c_s\Big(\sum_{j+\ell+k=0} |u_j|(|v_\ell||\ell|^{s+1}|w_k|) + \sum_{j+\ell+k=0} |v_\ell||\ell||u_j||j|^s|w_k|\Big)$$

$$\leq (2\pi)^{s+1}c_s\Big(\sum_{j\in\mathbb{Z}^n}|u_j|\Big)|w||A^{\frac{s+1}{2}}v| + (2\pi)^{s+1}c_s\Big(\sum_{\ell\in\mathbb{Z}^n}|v_\ell||\ell|\Big)(|w||A^{s/2}u|).$$

All we need now is to observe that $\sum_{j\in\mathbb{Z}^n} |u_j| \leq c|A^{s/2}u|$ and $\sum_{\ell\in\mathbb{Z}^n}|\ell||v_\ell| \leq c|A^{\frac{s+1}{2}}v|$. Indeed,

$$\sum |u_j| = \sum |j|^{-s}|j|^s|u_j| \leq \Big(\sum_j |j|^{-2s}\Big)^{1/2}\Big(\sum_j |j|^{2s}|u_j|^2\Big)^{1/2} \leq c|A^{s/2}u|$$

since $2s > n$. Also

$$\sum |\ell||v_\ell| = \sum_\ell |\ell|^{-s}|\ell|^{s+1}|v_\ell| \leq \Big(\sum_{\ell\in\mathbb{Z}^n\setminus 0}|\ell|^{-2s}\Big)^{1/2}\Big(\sum_{\ell\in\mathbb{Z}^n}|\ell|^{2s+2}|v_\ell|^2\Big)^{1/2}$$

$$\leq c|A^{\frac{s+1}{2}}v| .$$

This proves (10.8). Moreover, we note that (10.8) is true for $s > n/2$.

Now for the proof of (10.9). We have, since $(B(u,A^{s/2}v),A^{2/2}v) = 0$

$$(A^sB(u,v),v) = (A^{s/2}B(u,v) - B(u,A^{s/2}v),A^{s/2}v).$$

We will prove

$$(10.11) \quad |A^{s/2}B(u,v) - B(u,A^{s/2}v)| \le c|A^{s/2}u||A^{s/2}v|$$

Again we check (10.11) by estimating $\sup_{|w|=1} |(A^{s/2}B(u,v) - B(u,A^{s/2}v),w)|$.

$$(A^{s/2}B(u,v),w) = (B(u,v),A^{s/2}w) = (2\pi i)(2\pi)^s \sum_{j+k+\ell=0} \langle u_j,k\rangle \langle v_k,w_\ell\rangle |\ell|^s.$$

$$(B(u,A^{s/2}v),w) = (2\pi i)(2\pi)^s \sum_{j+k+\ell=0} \langle u_j,k\rangle \langle v_k,w_\ell\rangle |k|^s.$$

Taking the difference

$$(10.12) \quad (A^{s/2}B(u,v) - B(u,A^{s/2}v),w)$$

$$= (2\pi i)(2\pi)^s \sum_{\substack{j+k+\ell=0 \\ j,k,\ell \in Z^n\setminus 0}} \langle u_j,k\rangle \langle v_k,w_\ell\rangle(|\ell|^s - |k|^s).$$

Now, for any $s \ge 1$ there exists c_s such that

$$(10.13) \quad \Big||\xi|^s - |\eta|^s\Big| \le c_s|\xi - \eta|[|\xi - \eta|^{s-1} + |\eta|^{s-1}]$$

for any $\xi,\eta \in \mathbb{R}^n$.

Indeed,

$$|\xi|^s - |\eta|^s = \int_0^1 \frac{d}{dt} |t(\xi - \eta) + \eta|^s dt$$

and

$$\frac{d}{dt} |t(\xi - \eta) + \eta|^s = s|t(\xi - \eta) + \eta|^{s-1} \frac{\langle t(\xi - \eta) + \eta, \xi - \eta\rangle}{|t(\xi - \eta) + \eta|}.$$

Thus using $0 < t < 1$ and (10.10)

$$\left|\frac{d}{dt}|t(\xi - \eta) + \eta|^s\right| \le s|\xi - \eta||t(\xi - \eta) + \xi|^{s-1}$$

$$\le c_s|\xi - \eta|[|\xi - \nu|^{s-1} + |\eta|^{s-1}].$$

Now in (10.12), $|\ell| = |j + k|$ and using (10.13) with $\xi = j+k$, $\eta = k$ we get

$$|(A^{s/2}B(u,v) - B(u,A^{s/2}v),w)| \le$$

$$\le (2\pi)^{s+1}c_s \sum_{\substack{j+k+\ell=0 \\ j,k,\ell \in \mathbf{Z}^n\backslash 0}} |j||u_j||k||v_k||w_\ell|[|j|^{s-1} + |k|^{s-1}] =$$

$$= (2\pi)^{s+1}c_s \sum_{j+k+\ell=0} |j|^s|u_j||w_\ell||k||v_k|$$

$$+ (2\pi)^{s+1} c_s \sum_{j+k+\ell=0} |k|^s|v_k||w_\ell||j||u_j|.$$

Let us estimate the first sum:

$$\sum_{j+k+\ell=0} |j|^s|u_j||w_\ell||k||v_k| = \sum_k |k||v_k| \sum_j |j|^s|u_j||w_{-j-k}|$$

We get, thus

$$\sum_{j+k+\ell=0} |k||v_k||j|^s|u_j||w_\ell| \le c|A^{s/2}u||w| \sum_{k \in Z^n} |k||v_k|$$

Similarly,

$$\sum_{j+k+\ell=0} |j||u_j||k|^s|v_k||w_\ell| \le c\Big(\sum_{j \in \mathbf{Z}^n} |j||u_j|\Big)|A^{s/2}v||w|.$$

Now since $s > \frac{n}{2} + 1$ we have

$$\sum_{k \in Z^n\backslash 0} |k||v_k| = \sum_{k \in Z^n\backslash 0} |k|^s|v_k||k|^{1-s} \le \Big(\sum_{k \in Z^n\backslash 0} |k|^{2s}|v_k|^2\Big)^{1/2}\Big(\sum_{k \in Z^n\backslash 0} |k|^{2-2s}$$

$$\le \Big(\sum_{k \in Z^n\backslash 0} |k|^{2s}|v_k|^2\Big)^{1/2}\Big(\sum_{k \in Z^n\backslash 0} |k|^{2-2s}\Big)^{1/2} \le c|A^{s/2}v|.$$

This means

(10.14) $\sum_k |k||v_k| \leq c|A^{s/2}v|, \quad s > \frac{n}{2} + 1$

The proof of (10.11) and therefore of (10.9) is complete.

Remark 10.5. Estimates of the type of (10.8), (10.9), and (10.11), suitably modified, are valid for $\mathbb{R}^n$ replacing T^n, the unit torus.

Now we are going to make more precise the statement about higher regularity of strong solutions. Let us return to n = 3 and assume that f is (for simplicity) time independent and u_0,f, are smooth. Let us consider the periodic case and fix $\nu > 0$, $L > 0$. Let us assume that u(t) solution of (10.1), (10.2) is a strong solution on [0,T]: $u \in L^\infty(0,T;V) \cap L^2(0,T;\mathcal{D}(A))$. Let us first assume that $f, u_0 \in H^s(T^3)^3$ with $s > 3/2$. Since the estimate (10.8) is valid for $s > n/2$ we get from (10.1), multiplying scalarly by $A^s u$:

$$\frac{1}{2}\frac{d}{dt}|A^{s/2}u|^2 + \nu|A^{\frac{s+1}{2}}u|^2 \leq |A^{s/2}f||A^{s/2}u| + cL^{s-\frac{3}{2}}|A^{\frac{s+1}{2}}u||A^{s/2}u|.$$

Using by now familiar manipulations we deduce that

$$\frac{d}{dt}|A^{s/2}u|^2 + \nu|A^{\frac{s+1}{2}}u|^2 \leq \frac{c}{\nu}[L^{2s-3}|A^{s/2}u|^4 + |A^{s/2}f|^2].$$

Thus by Gronwall's inequality

(10.15) $u \in L^\infty(0,T;\mathcal{D}(A^{s/2}) \cap L^2(0,T;\mathcal{D}(A^{\frac{s+1}{2}}))$

provided $u \in L^2(0,T;\mathcal{D}(A^{s/2}))$. If $3/2 < s \leq 2$ then $u \in L^2(0,T:\mathcal{D}(A)) \cap L^2(0,T;\mathcal{D}(A^{s/2}))$. It follows that, if u is a strong solution with u_0, $f \in H^s(T^3)^3$ for $s \in (3/2,2]$ then $u \in L^\infty(0,T;\mathcal{D}(A^{s/2})) \cap L^2(0,T;\mathcal{D}(A^{\frac{s+1}{2}}))$.

In particular u(t) is bounded in H^s. Now, if $s \in (2,5/2]$ then $s' = s - \frac{1}{2} \in (\frac{3}{2},2]$ and therefore a strong solution $u \in L^2(0,T;\mathcal{D}(A^{\frac{s+1}{2}}))$. Since $\frac{s'+1}{2} = \frac{s}{2} + \frac{1}{4} > \frac{s}{2}$ it follows from (10.15) that $u \in L^\infty(0,T;\mathcal{D}(A^{s/2})) \cap L^2(0,T;\mathcal{D}(A^{\frac{s+1}{2}}))$ for $s \in (2,5/2]$. We can go on like this forever.

These were formal considerations but can be easily made rigorous.

Theorem 10.6. Let $s > 3/2$. Assume $f,u_0 \in \mathcal{D}(A^{s/2}) = V \cap H^s(T^3)^3$. Assume that u(t), solution of (10.1), (10.2) (periodic case) is a strong solution, i.e., $u \in L^\infty(0,T;V) \cap L^2(0,T;\mathcal{D}(A))$. Then $u \in L^\infty(0,T;\mathcal{D}(A^{s/2})) \cap L^2(0,T;\mathcal{D}(A^{\frac{s+1}{2}}))$.

Corollary 10.7. Let $s > 3/2$. Assume $f,u_0 \in \mathcal{D}(A^{s/2})$. Then a necessary and sufficient conditions for a weak solution $u \in L^\infty(0,T;H) \cap L^2(0,T;V)$ of (10.1), (10.2) (periodic case) to belong to $L^\infty(0,T;\mathcal{D}(A^{s/2})) \cap L^2(0,T;\mathcal{D}(A^{\frac{1+s}{2}}))$ is that $u \in L^4(0,T;\mathcal{D}(A^{1/2}))$.

Proof. Clearly, if $s > 3/2$,

$$L^\infty(0,T;\mathcal{D}(A^{s/2})) \cap L^\infty(0,T;\mathcal{D}(A^{1/2})) \subset L^4(0,T;\mathcal{D}(A^{1/2})).$$

Reciprocally, if $u \in L^4(0,T;\mathcal{D}(A^{1/2}))$ and $u_0,f \in \mathcal{D}(A^{s/2})$, $s > 3/2$ then formally

$$(10.16) \qquad \frac{d}{dt}\|u\|^2 + \nu|Au|^2 \le \frac{2|f|^2}{\nu} + \frac{c}{\nu^3}\|u\|^4\|u\|^2.$$

Since $\int_0^T \|u\|^4 dt < \infty$ we deduce that u is a strong solution on [0,T]. The rest follows from Theorem 10.6. To make this into a bona fide proof we observe that, because of the local existence theorem (10.16) is true for

$t \leq T_0$. We use the local existence and uniqueness theorem to prove $\sup\{T_0 > 0 | u \text{ is a strong solution on } [0,T_0]\} \geq T$. We omit further details.

We presented Theorem 10.6 and Corollary 10.7 as justifications of the importance of the question: do strong solutions lose regularity? We do not, as yet, know the answer to this question. We saw that loss of regularity occurs only if $\|u(t)\|$ becomes infinite. The set of singular times of a weak solution is clearly of Lebesgue measure 0 since $\int_0^T \|u(t)\|^2 dt$ is finite. Actually more is true: The set of singular times of a weak solution has Hausdorff dimension not larger than 1/2.

Let us assume that $u_0 \in V$ and $f \in H$ (for simplicity). Let $u(t)$ be a weak solution of (10.1), (10.2). Assume that $u(t_0) \in V$ for some t_0. Then, the local existence theorem (Theorem 9.4) and the uniqueness result (Remark 10.3 a)) imply that $u|_{[t_0,t_0+T_0]}$ is a regular solution (i.e., $u \in L^\infty(t_0,t_0+T_0;V) \cap L^2(t_0,t_0+T_0,\mathcal{D}(A))$ for some positive T_0 depending on $\|u(t_0)\|$. Actually, from (9.32) it follows that

$$\nu\lambda_1 T_0\left(1 + \frac{|f|^2}{\nu^4 \lambda_1^{3/2}}\right) \geq c\left(1 + \frac{\|u(t_0)\|^2}{\nu^2 \lambda_1^{1/2}}\right)^{-2}. \tag{10.17}$$

For each t_0 such that $\|u(t_0)\|$ is finite we consider a maximal interval $I \subset [0,T]$ on which u is regular. More precisely, I maximal with the properties: I is an interval included in $[0,T]$, $t_0 \in I$, and for any closed interval $J \subset I$, $u|_J \in L^\infty(J,V) \cap L^2(J,\mathcal{D}(A))$. The existence of a maximal interval with these properties follows from the fact that the set of intervals with these properties is nonempty and is inductively ordered under inclusion. I is necessarily open at the right end if the right end is not T. Clearly there are at most countably many distinct intervals of this type, I_j, and the measure of $[0,T] \setminus \bigcup_{j=1}^\infty I_j$ is zero. (This is obvious since $\int_0^T \|u(t)\|^2 < \infty$ and thus $\|u(t_0)\| < \infty$ for almost all t_0).

Let I be one of the intervals I_j. From (10.17) it follows that for any $t_0 \in I$

$$\nu\lambda_1(b - t_0)\left(1 + \frac{|f|^2}{\nu^4\lambda_1^{3/2}}\right) \geq c\left(1 + \frac{\|u(t_0)\|^2}{\nu^2\lambda_1^{1/2}}\right)^{-2}$$

if b is the least upper bound of I. It follows that

$$(10.18) \quad \frac{1}{\sqrt{b - t_0}} \leq \left(1 + \frac{\|u(t_0)\|^2}{\nu^2\lambda_1^{1/2}}\right)(\nu\lambda_1)^{1/2}\left(1 + \frac{|f|^2}{\nu^4\lambda_1^{3/2}}\right)^{1/2}$$

Integrating (10.18) on I we get

$$2|I_j|^{1/2} \leq K\left(|I_j| + \frac{1}{\nu^2\lambda_1^{1/2}}\int_{I_j} \|u(t_0)\|^2 dt_0\right)$$

Summing over j we obtain

$$(10.19) \quad \sum_j |I_j|^{1/2} \leq K\left(T + \frac{1}{\nu^2\lambda_1^{1/2}}\int_0^T \|u\|^2 dt\right) < \infty.$$

Let $X \subset M$ be a compact subset of a metric space. We define the d-dimensional (outer) Hausdorff measure of X by

$$(10.20) \quad \mu_H^d(X) = \lim_{r \to 0} \mu_{H,r}^d(X)$$

where

$$(10.21) \quad \mu_{H,r}^d(X) = \inf\left\{\sum_{i=1}^k r_i^d \mid X \subset \bigcup_{i=1}^k B_i,\ B_i \text{ open balls in } M \text{ of radius } r_i \leq r\right\}$$

The Hausdorff dimension of X is

$$d_H(X) = \inf\{d > 0 \mid \mu_H^d(X) = 0\}.$$

Iet us consider a weak solution u(t) defined on [0,T] of (10.1), (10.2).

Let $\{I_j\}_{j \in \mathbb{N}}$ be the collection of maximal intervals of regularity in $[0,T]$ described earlier. Let us denote by a_j and b_j their ends:

$$(a_j,b_j) \subset I_j \subset [a_j,b_j).$$

Clearly a_j and b_j are singular times, in particular, $\|u(b_j)\| = \infty$ (if you prefer $\limsup_{t \to b_j} \|u(t)\| = +\infty$).

Theorem 10.8. Let u be a weak solution of (10.1), (10.2) on $[0,T]$. There exists a set E, closed, of 1/2-dimensional Hausdorff measure 0, outside which u is regular, i.e.,

$$u|_{[0,T]\setminus E} \in L^\infty_{loc}((0,T)\setminus E;V) \cap L^2_{loc}((0,T)\setminus E \cap \mathcal{D}(A)).$$

Proof. E will be defined as $E = [0,T] \setminus \bigcup_{j=1}^{\infty} \overset{\circ}{I}_j$ where I_j on the maximal intervals of regularity constructed above and $\overset{\circ}{I}$ means interior of I. In order to compute the 1/2-dimensional Hausdorff measure of E, let us first make the observation that in the definition of μ_H^d we can use closed intervals instead of open intervals in the case the metric space M is $\mathbb{R}$. Let then m be a positive integer. Let E_m be the set $E_m = [0,T] \setminus \bigcup_{j=1}^{m} \overset{\circ}{I}_j$; $E_m \supset E$. Clearly E_m is the union of a finite number of closed intervals $E_m = \bigcup_{j=1}^{k_m} K_j^{(m)}$. The intervals $K_j^{(m)}$ are closed and disjoint. Since the Lebesgue measure of $\bigcup_{j=1}^{\infty} I_j$ is T it follows that the measure of any of the $K_m^{(\ell)}$'s (i.e, its length) is not larger than the sum of the measure of the I_j that touch it. From the construction of the $K_j^{(m)}$'s if an inteval I_j with $j \geq m+1$ touches a $K_\ell^{(m)}$ then it is included in it. Thus, the sets $N_\ell^{(m)} = \{j \geq m+1 ; I_j \cap K_\ell^{(m)} \neq \emptyset\} = \{j \geq m+1 ; I_j \subset K_\ell^{(m)}\}$ are disjoint. Denoting by $|F|$ = the Lebesgue measure of F we have

$$|K_\ell^{(m)}| \leq \sum_{j \in N_\ell^{(m)}} |I_j|$$

Thus

$$|K_\ell^{(m)}| \leq \sum_{j=m+1}^{\infty} |I_j| = \varepsilon_m$$

and also

$$|K_\ell^{(m)}|^{1/2} \leq \sum_{j \in N_\ell^{(m)}} |I_j|^{1/2}.$$

Then

$$\sum_{\ell=1}^{k_m} |K_\ell^{(m)}|^{1/2} \leq \sum_{j \geq m+1} |I_j|^{1/2} \xrightarrow[m \to \infty]{} 0.$$

Since $\bigcup_{\ell=1}^{k_m} K_\ell^{(m)}$ is a cover of E with intervals of radius less than $\varepsilon_m/2$, we proved here that

$$\mu^{1/2}_{H,\varepsilon_m/2}(E) \leq \delta_m = \sum_{j=m+1}^{\infty} |I_j|^{1/2}.$$

In view of (10.19), $\delta_m \to 0$ and thus $\mu_H^{1/2}(E) = 0$.

The next result shows that if the enstrophy $\|u(t)\|^2$ of a solution of the three-dimensional Navier-Stokes equations (10.1) becomes infinite in infinite time then there are solutions $\tilde{u}(t)$ of the same equations for which the enstrophy becomes infinite in finite time.

Theorem 10.9. Let $u(t)$ be a solution of the three-dimensional Navier-Stokes equations (10.1) in a domain $\Omega \subset \mathbb{R}^3$ with C^2 boundary. Assume $f \in H$ is time independent (for simplicity). Assume $u(t)$ is a strong soution for each $T > 0$; $u(0) = u_0 \in V$; $u \in L^\infty(0,T;V) \cap L^2(0,T; \mathcal{D}(A))$ for all $T > 0$. Assume also that $\limsup_{t \to \infty} \|u(t)\| = +\infty$. Then, for any $T_1 > 0$ there exists $v_0 \in V$ such that the solution to (10.1) having v_0 as initial data blows up before $T_1 > 0$, i.e., v is not a stong solution on $[0,T_1]$.

Proof. We use (9.11) which is valid for 3-dimensional Navier-Stokes too. Let $t_j \to \infty$ be such that $\lim_{j\to\infty} \|u(t_j)\| = \infty$. From (9.11) we find $a_j \in [t_j - T_1, t_j]$ such that

$$\|u(a_j)\|^2 \leq \frac{K_1}{T_1} + K_2 = K$$

for appropriate constants K_1, K_2, K which are j independent.

Passing to a subsequence we may assume that $u(a_j)$ converges weakly in V and strongly in H to an element v_0 of V. Now let us consider the functions $v_j(s) = u_j(s + a_j)$ and the function $v(s)$ defined by solving the Navier-Stokes equation (10.1) with initial data $v_0 \in V$. By the local existence theorem $v(s)$ is a strong solution for some $T_0 = T_0(\|v_0\|)$.

Our claim is that v cannot be a strong solution on $[0,T_1]$. Indeed, assume by contradiction that u is a strong solution. Forming the differences $w_j(s) = v_j(s) - v(s)$ we have

$$\frac{d}{dt} w_j + \nu A w_j + B(w_j, v) + B(v, w_j) + B(w_j, w_j) = 0$$

$$w_j(0) = v_j(0) - v_0.$$

Taking the scalar product of the first equation with $w_j(s)$, we get

$$\frac{1}{2}\frac{d}{dt} |w_j(t)|^2 + \nu \|w_j(t)\|^2 \leq c\|v\| \, \|w_j\|^{3/2} |w_j|^{1/2}$$

$$\leq \frac{\nu}{2} \|w_j\|^2 + \frac{c}{\nu^3} \|v\|^4 |w_j|^2$$

From Gronwall

$$|w_j(t)|^2 \leq |w_j(0)|^2 \exp \int_0^t \frac{c}{\nu^3} \|v\|^4 ds$$

Since by assumption $\int_0^{T_1} \|v\|^4 ds$ is finite and $v_j(0) - v_0 \to 0$ in H it follows that $w_j(t) \to 0$ in H for all $t \in [0,T_1]$. Also since

$$|w_j(t)|^2 + \nu \int_0^t \|w_j(s)\|^2 ds \leq \frac{c}{\nu^3} \int_0^t \|v\|^4 |w_j(s)|^2$$

it follows that $\int_0^T \|w_j(s)\|^2 ds$ tends to zero. Thus $\|w_j(t)\| \to 0$ a.e. for $t \in [0,T_1]$. Now

$$\|v_j(t)\| \le \|w_j(t)\| + \|v(t)\|$$

$$\le \|w_j(t)\| + \|v\|_{L^\infty(0,T_1;V)} = \|w_j(t)\| + r.$$

Take any $t \in [0,T_1]$ such that $\|w_j(t)\| \le 1$. Then $\|v_j(t)\| \le 1+r$. By the local existence theorem , (Theorem 9.4) there exists a time interval $[t,t+T_2]$ with T_2 depending on r and $|f|$, ν, λ_1 only, <u>but independent of t</u> such that

$$\|v_j(s)\|^2 \le \sqrt{2}\ (\|v_j(t)\|^2 + 1) \le \sqrt{2}((1 + r)^2 + 1)$$

for all $s \in [t,t+T_2]$.

Let I be the set of $t \in [0,T_1]$ such that $\lim_{j \to \infty} \|w_j(t)\| = 0$. Clearly there exists finitely many $t_i \in I$, $i = 1,\dots,m$ such that $[T_0,T_1] \subset \bigcup_{i=1}^m [t_i,t_i+ T_2]$ where T_2 is defined above. Therefore, there exists $j_0 \ge 1$ such that for $t \in [0,T_1]$, $j \ge j_0$, $\|v_j(t)\| \le c$, absurd, since

$$\|v_j(t_j - a_j)\| = \|u(t_j)\| \to \infty.$$

11

VANISHING VISCOSITY LIMITS

Let us consider solutions of

(11.1) $\frac{du}{dt} + \nu Au + B(u,u) = f$

(11.2) $u(0) = u_0$

where the problem is set in the periodic n-dimenisonal case (n = 2,3). The function f is, for simplicity, assumed to be time independent. We recall that the period of the functions u(t,.) is denoted by L. Let us assume that $f \in \mathcal{D}(A^{s/2})$ and $u_0 \in \mathcal{D}(A^{s/2})$ for some $s > 1 + \frac{n}{2}$. Let us multiply (11.1) by $A^s u$ and use (10.9):

(11.3) $\frac{1}{2}\frac{d}{dt}|A^{s/2}u|^2 + \nu|A^{\frac{s+1}{2}}u|^2 \leq |A^{s/2}f||A^{s/2}u| + cL^{s-1-\frac{n}{2}}|A^{s/2}u|^3$.

Let us define the scale independent quantities

(11.4) $\tau = \frac{\nu}{L^2} t$

(11.5) $y_s(\tau) = \frac{L^{s-\frac{n}{2}+1}}{\nu} |A^{s/2}u(\frac{L^2\tau}{\nu})|$

(11.6) $g_s = \frac{L^{s-\frac{n}{2}+3}}{\nu^2} |A^{s/2}f|$.

Multiplying (11.3) by L^{2s-n+4}/ν^3 we get

(11.7) $\frac{d}{d\tau} y_s(\tau) \leq g_s + c(y_s(\tau))^2$

(In order to obtain (11.7) one neglects the nonnegative term $\nu|A^{\frac{s+1}{2}}u|^2 \cdot \frac{L^{2s-n+4}}{\nu^3}$ and divides by $\sqrt{y}$.) Let us drop, for simplicity, the index s. We infer from (11.7) that

$$(11.8) \qquad \frac{\frac{dy}{d\tau}}{g + cy^2} \leq 1.$$

Integrating between 0 and τ we obtain

$$\arctan\sqrt{\frac{c}{g}}\, y(\tau) - \arctan\sqrt{\frac{c}{g}}\, y(0) \leq \sqrt{cg}\ \tau.$$

Therfore, if

$$(11.9) \qquad \sqrt{cg}\ \tau + \arctan\sqrt{\frac{c}{g}}\, y(0) < \frac{\pi}{2}$$

we obtain

$$\sqrt{\frac{c}{g}}\, y(\tau) \leq \tan(\sqrt{cg}\ \tau + \arctan\sqrt{\frac{c}{g}}\, y(0)).$$

Using a well known trigonometric formula

$$(11.10) \qquad \sqrt{\frac{c}{g}}\, y(\tau) \leq \frac{\tan(\sqrt{cg}\ \tau) + \sqrt{\frac{c}{g}}\, y(0)}{1 - \tan(\sqrt{cg}\ \tau)\sqrt{\frac{c}{g}}\, y(0)}.$$

If $y = 0$ we get, directly from (11.8) that

$$(11.11) \qquad y(\tau) \leq \frac{y(0)}{1 - c\tau y(0)}$$

We get, in the $g = 0$ case, if

$$(11.12) \qquad \tau y(0)c \leq 1/2$$

that

$$(11.13) \qquad y(\tau) \leq 2y(0).$$

This means, going back to the scale dependent quantities that, for $f = 0$,

$$(11.14)\quad |A^{s/2}u(t)| \le 2|A^{s/2}u(0)|$$

provided

$$(11.15)\quad cL^{s-\frac{n}{2}-1}|A^{s/2}u(0)|t \le \frac{1}{2}\,.$$

Note that condition (11.15) is independent of ν. If $g \neq 0$ then we find $\varepsilon > 0$ such that, if

$$(11.16)\quad cy(0)\tau \le 1/4$$

$$(11.17)\quad \sqrt{cg}\;\tau \le \varepsilon$$

then

$$(11.18)\quad y(\tau) \le 2(2\tau g + y(0)).$$

(The number ε is determined here by the requirement $\tan x < 2x$ for $0 < x < \varepsilon$.) Going back to scale dependent quantities, we get that

$$(11.19)\quad |A^{s/2}u(t)| \le 4t|A^{s/2}f| + 2|A^{s/2}u_0|$$

provided that

$$(11.20)\quad t \le \mathrm{Min}\{\frac{1}{4c}\,|A^{s/2}u_0|^{-1}L^{1+\frac{n}{2}-s}\,;\ \frac{\varepsilon}{\sqrt{c}}\,|A^{s/2}f|^{-1/2}L^{\frac{1}{2}+\frac{n}{4}-\frac{s}{2}}\}.$$

Estimates of the type (11.19) and the vanishing viscosity limit were obtained for the $L = \infty$ case in ([K1]).

<u>Theorem 11.1</u>. Let $f \in \mathcal{D}(A^{s/2})$ and $u_0 \in \mathcal{D}(A^{s/2})$ for $s > 1 + \frac{n}{2}$ and A defined in the L-periodic, n-dimensional case. There exists $T > 0$ depend

ing on L, s, u_0, and f (T given by the right-hand side of (11.20)) such that for every $\nu > 0$ and corresponding solution $u(t) = u_\nu(t)$ of (11.1) and (11.2), the bound (11.19) is valid.

Moreover, if $s > 3 + \frac{n}{2}$ then as ν converges to 0 the functions u_ν converge uniformly in $L^\infty(0,T;\mathcal{D}(A^{s'/2}))$, $1 + \frac{n}{2} s' \leq s-2$, to a function v, a solution of the Euler equations

$$(11.21) \quad \frac{dv}{dt} + B(v,v) = f$$

$$(11.22) \quad v(0) = u_0$$

Proof Let us take two solutions $u_\nu(t)$ and $u_\mu(t)$ of (11.1), (11.2). Let us assume $\nu > \mu$. Then forming the difference $w(t) = u_\nu(t) - u_\mu(t)$ we obtain the equation

$$\begin{cases} \frac{dw}{dt} + \nu Aw + B(w,u_\mu) + B(u_\mu,w) + B(w,w) = (\mu - \nu)Au_\mu, \\ \\ w(0) = 0 \end{cases}$$

Taking the scalar product with $A^{s'/2}w$ and using (10.8), (10.9) we obtain

$$\frac{1}{2}\frac{d}{dt}|A^{\frac{s'}{2}} w|^2 \leq k[|A^{\frac{s'+1}{2}} u_\mu||A^{\frac{s'}{2}} w|^2 + |A^{\frac{s'}{2}} u_\mu||A^{\frac{s'}{2}} w|^2 + |A^{\frac{s'}{2}} w|^3]$$
$$+ |\mu - \nu||A^{\frac{s'}{2}+1} u_\mu||A^{\frac{s'}{2}}w|.$$

The constant k depends on L, s'. Now since $s'+2 \leq s$ we can bound $|A^{\frac{s'}{2}+1} u_\mu|$ and thus $|A^{\frac{s'}{2}} u_\mu|$, $|A^{\frac{s'}{2}+\frac{1}{2}} u_\mu|$ uniformly on [0,T], using (11.19). Dividing by $|A^{s'/2}w|$ we get for $y = |A^{s'/2}w|$ an inequality of the type

$$\frac{dy}{dt} \leq K(y + y^2 + \nu)$$

for $t \in [0,T]$. Multiplying by e^{-Kt} and considering $z = e^{-Kt}y$ we get

$$(11.23)\qquad \begin{cases} \dfrac{dz}{dt} \le Mz^2 + M\nu \\ z(0) = 0 \end{cases}$$

for $t \le T$, and some M depending on L,T. This inequality is of the same type as (11.7). Therefore, if ν is small (see (11.17)) i.e., if

$$(11.24)\qquad \nu \le \frac{\varepsilon^2}{M^2T^2}$$

then from (11.18) we deduce

$$(11.25)\quad z(t) \le 4TM\nu$$

for all $t \in [0,T]$.

Form (11.25) we deduce that

$$(11.26)\qquad |A^{s'/2}(u_\nu - u_\mu)(t)| \le (4TMe^{KT})\nu$$

for all $t \le T$ and $\nu \le \varepsilon^2/(M^2T^2)$, $\mu < \nu$.

The inequality (11.26) enables us to pass to the limit in (10.1). We omit further details.

The same technique is used in [C1] to prove that if u_0 is smooth enough, and if the solution v(t) to the Euler equations (11.21), (11.22) is smooth on an interval $[0,T_1]$ then the solutions u_ν of (11.1), (11.2) will be smooth on the same interval $[0,T_1]$ for all $0 < \nu \le \nu_0$ where ν_0 is a positive number determined by the solution v(t).

12
ANALYTICITY AND BACKWARD UNIQUENESS

We discuss first analyticity of stong solutions as $\mathcal{D}(A)$ valued functions of time ([T.2]). Let us consider a Galerkin system

$$(12.1) \quad \begin{cases} \dfrac{du_m}{dt} + \nu A u_m + P_m B(u_m, u_m) = P_m f \\ u_m(0) = P_m u_0 \end{cases}$$

where $f \in H$, $u_0 \in V$, $m \geq 1$ and the system originates from any of the $n = 2,3$ Dirichlet or periodic cases.

We want to extend (12.1) for complex t. In order to do so we need to complexify the spaces $H, H_m, V, \mathcal{D}(A)$ and the corresponding operators. For instance, the complexification of H is

$$H_{\mathbf{C}} = \{u_1 + iu_2 | \ u_1 \in H, \ u_2 \in H\}.$$

The scalar product will be

$$(u,v)_{\mathbf{C}} = (u_1 + iu_2, v_1 + iv_2)_{\mathbf{C}} = (u_1,v_1) + (u_2,v_2) + i[(u_2,v_1) - (u_1,v_2)]$$

The system (12.1) admits a unique solution $u_m(t)$ for t in a complex neighborhood of the origin. Since f and u_0 are real, the solution $u_m(t)$, for t real, is real and coincides with the usual Galerkin approximation. We want to prove that the complex domain of definition of $u_m(t)$ can be chosen independently of m and that a a priori bounds can assure passage to the limit in $m \to \infty$.

Let us fix $\theta \in (-\frac{\pi}{2}, \frac{\pi}{2})$ and take t of the form $t = se^{i\theta}$ for $s > 0$. We want to compute

$$\frac{d}{ds} \| u_m(se^{i\theta}) \|^2 = \frac{d}{ds} (u_m(se^{i\theta}), Au_m(se^{i\theta})).$$

We get

$$\frac{1}{2}\frac{d}{ds} \| u_m(se^{i\theta}) \|^2 = \frac{1}{2} (e^{i\theta} \frac{du_m}{dt}, Au_m) + \frac{1}{2} (u_m, e^{i\theta} A \frac{du_m}{dt})$$

$$= \operatorname{Re} e^{i\theta}(\frac{du_m}{dt}, Au_m).$$

Thus, multiplying (12.1) scarlarly by $Au_m(t)$ and by $e^{i\theta}$ and taking the real part we obtain

$$(12.2) \qquad \frac{1}{2}\frac{d}{ds} \| u_m(se^{i\theta}) \|^2 + \nu \cos\theta \, |Au_m(se^{i\theta})|^2$$

$$= \operatorname{Re}\{e^{i\theta}[(B(u_m,u_m),Au_m)_{\mathbf{C}} + (f,Au_m)_{\mathbf{C}}]\}.$$

Now we use the estimates

$$(12.3) \qquad |(B(u_m,u_m), Au_m)_{\mathbf{C}}| \le C\|u_m\|^{3/2}|Au_m|^{3/2}$$

which follow from corresponding estimates for the real case.

The term involving $(f,Au_m)_{\mathbf{C}}$ will be estimated

$$|(f,Au_m)| \le \frac{\nu \cos\theta}{4} |Au_m|^2 + \frac{|f|^2}{\nu \cos\theta}.$$

Also from (12.3) we get

$$(12.4) \qquad |(B(u_m,u_m,Au_m)_{\mathbf{C}}| \le \frac{\nu \cos\theta}{4} |Au_m|^2 + \frac{C}{\nu^3(\cos\theta)^3} \|u_m\|^6$$

We deduce the inequality

$$(12.5) \qquad \frac{d}{ds} \| u_m(se^{i\theta}) \|^2 + \nu \cos\theta |Au_m(se^{i\theta})|^2$$

$$\le \frac{2|f|^2}{\nu \cos\theta} + \frac{C}{\nu^3(\cos\theta)^3} \| u_m(se^{i\theta}) \|^6.$$

The inequality (12.5) is valid for all the cases ($n = 2,3$, Dirichlet or periodic). We see that on a time ray of fixed angle θ the role of the viscosity is played by $\nu \cos \theta$.

We obtain a bound for $\|u_m(t)\|^2$, $t = se^{i\theta}$ of the type

$$(12.6) \qquad \|u_m(t)\|^2 \leq \sqrt{2}(\|u_0\|^2 + 1)$$

provided $t = se^{i\theta}$ satisfies

$$(12.7) \qquad s\left(\frac{4|f|^2}{\nu \cos \theta} + \frac{2C}{\nu^3(\cos \theta)^3}\right) \leq \frac{1}{2(1 + \|u_0\|^2)^2}$$

(see (9.31) and the proof of Theorem 9.4). (The inequality (12.7) is not nondimensional.) Fixing the parameters $|f|, \nu, \|u_0\|$, we see that (12.6) amounts to a uniform bound for $\|u_m(t)\|$ for all $m \geq 1$ and t in a region $D = D(\nu, \|u_0\|, |f|) \subset \mathbb{C}$ described in (12.7).

The open set

$$D = D(\nu, \|u_0\|, |f|) = \left\{ t = se^{i\theta} \;\middle|\; |\theta| < \pi/2, \right.$$
$$\left. 0 < s\left(\frac{4|f|^2}{\nu \cos \theta} + \frac{2C}{\nu^3(\cos \theta)^3}\right) < \frac{1}{2(1 + \|u_0\|^2)^2} \right\}$$

is therefore a domain of analyticity of the functions $u_m(t)$. It is symmetric about the real axis and thus is a neighborhood of all its real points. The origin 0 belongs to its closure.

In order to obtain a priori bounds for $|Au_m(t)|$ for $t \in D$ we use first the Cauchy formula to obtain a priori bounds for $\|\frac{du_m}{dt}\|$. Indeed, let γ be a small circle contained in D. Then, for t inside the circle

$$(12.8) \qquad \frac{d^k u_m}{dt^k}(t) = \frac{k!}{2\pi i} \int_\gamma \frac{u_m(z)}{(z - t)^{k+1}}\, dz$$

and therefore

$$(12.9) \qquad \left\| \frac{d^k u_m}{dt^k}(t) \right\| \leq \frac{k!}{(r_\gamma)^k} \cdot 2^{1/4}(1 + \|u_0\|^2)^{1/2}$$

Here r_γ is the radius of γ and we used the estimate (12.6) for $\|u_m(z)\|$, $z \in \gamma$.

If M is a compact subset of D then denoting by $r_M = \text{dist}(M, \partial D)$ we obtain from (12.9) the estimates

$$(12.10) \quad \left\| \frac{d^k u_m}{dt^k}(t) \right\| \le \frac{k!}{(r_M)^k} \cdot 2^{1/4}(1 + \|u_0\|^2)^{1/2}$$

for all $t \in M \subset\subset D$ and $k = 0,1,2,\ldots$. In particular, taking $k = 1$ and using the equation (12.1) we deduce that

$$(12.11) \quad |Au_m(t)| \le E$$

for all $t \in M$, $m \ge 1$. The positive number E depends on ν, $|f|$, $\|u_0\|$, the set M, but not on $t \in M$, nor m.

The proof of (12.11) is straightforward; one uses the equation (12.1), the estimate $|B(u_m,u_m)| \le c\|u_m\|^{3/2}|Au_m|^{1/2}$ and (12.10). Now we can use (12.11) instead of (12.6) in the estimate of the Cauchy integral (12.8). We obtain, for every compact set L a constant E_L such that

$$(12.12) \quad \left|A \frac{d^k u_m}{dt^k}(t)\right| \le E_L \frac{k!}{r^k}$$

for all $t \in L$, $k \ge 0$, $m \ge 1$. The constant r is smaller than the distance from L to ∂D and E_L is the constant E of (12.11) for a set M such that

$$\{z \in \mathbf{C} \mid \text{dist}(z,L) \le r\} = M \subset\subset D$$

We can pass to the limit in m. We observe that, since the domain $D(\nu, \|u_0\|, |f|)$ depends on the size of $\|u_0\|$ we can repeat this construction with $t_0 \in R_+$ instead of 0 as vertex. We obtain

<u>Theorem 12.1 (Time analyticity)</u>. Let $f \in H$, $u_0 \in V$, $\nu > 0$. Let $\Omega \subset \mathbb{R}^n$ be an open bounded set with $\partial\Omega$ of class C^2. Then

(i) If $n = 2$, there exists an open neighborhood D of $(0,\infty)$ in $\mathbb{C}$ such that the solution $u(t)$ of

$$(12.13)\qquad \begin{cases} \frac{du}{dt} + \nu\, Au + B(u,u) = f \\ \\ u(0) = u_0 \end{cases}$$

is analytic $u: D \to \mathcal{D}(A)$.

(ii) If $n = 3$ there exists $T_0 > 0$ and an open neighborhood in $\mathbb{C}$ of $(0,T_0)$, D_{T_0}, such that the solution of (12.13) is analytic $u: D_{T_0} \to \mathcal{D}(A)$.

Remarks. 1. The same result holds in the periodic case.

2. In the $n = 2$ case, we can allow $u_0 \in H$.

We are going to use this result in order to deduce backward uniqueness of strong solutions.

Theorem 12.2 (Backward uniqueness) Let u_1, u_2 be two strong solutions of the Navier-Stokes system

$$(12.14)\qquad \frac{du}{dt} + \nu Au + B(u,u) = f$$

(i) The two-dimensional case: We assume that f, $u_1(0)$, $u_2(0)$ are in H. Suppose there exists $t_0 \geq 0$ such that $u_1(t_0) = u_2(t_0)$. Then $u_1(t) = u_2(t)$ for all $t \geq 0$.

(ii) The three-dimensional case: We assume that $f \in H$, $u_1(0)$, $u_2(0)$ belong to V. There exists $T_0 > 0$ (a common time of existence and analyticity of strong solutions) such that, if, for some $t_0 \in [0,T_0)$, $u_1(t_0) = u_2(t_0)$ then it follows that $u_1(t) = u_2(t)$ for all $t \in [0,T_0)$.

Proof. From Theorem 12.1 (and the Remark following it in the case $u_1(0)$, $u_2(0)$ belonging only to H) the functions $u_1(t)$, $u_2(t)$ are analytic in

$(0,T_0)$ for the $n = 3$ case and in $(0,\infty)$ for the $n = 2$ case. If $u_1(t_0) = u_2(t_0)$ then from the uniqueness of strong solutions to (12.14) it follows that $u_1(t) = u_2(t)$ for $t \geq t_0$. From the analyticity, $u_1(t) = u_2(t)$ for all $t > 0$. But $u_1(t)$, $u_2(t)$ tend strongly in H to $u_1(0)$, $u_2(0)$ as $t \downarrow 0$. Thus, $u_1(0) = u_2(0)$ must occur, too.

13

EXPONENTIAL DECAY OF VOLUME ELEMENTS

Let u(t) be a solution of the Navier-Stokes equations

$$(13.1) \qquad \frac{du}{dt} + \nu Au + B(u,u) = f$$

$$(13.2) \qquad u(0) = u_0$$

Suppose that the initial data depends on a parameter $\alpha \in \mathbb{R}$. Differentiating (13.1) with respect to this parameter we obtain the equations governing the time evolution of infinitesimal displacements:

$$(13.3) \qquad \frac{dv}{dt} + \nu Av + B(u,v) + B(v,u) = 0$$

$$(13.4) \qquad v(0) = v_0$$

In (13.3) the function u(t) is playing the role of a known coefficient in the equation. Standard energy methods together with the *a priori* bounds for u(t) described in previous sections will provide estimates for the solutions of (13.3).

To fix ideas we will consider the two-dimensional Navier-Stokes equations. The concepts that we are going to define and study in this and the next sections are quite general. The common theme of these last sections is the study of long time behavior of solutions to (13.1) or similar equations. Let us denote by $S(t)u_0 = u(t)$ the solutions to

(13.1), (13.2). We view $S(t)$ as a map from H into H. The following are known properties of $S(t)$:

Proposition 13.1.

(i) $S(t+s)u_0 = S(t)S(s)u_0 \qquad t,s \geq 0,\ u_0 \in H$

(ii) $\lim_{t \downarrow 0} S(t)u_0 = u_0$ in H

(iii) $S:(0,\infty) \to \mathcal{D}(A)$ is analytic

(iv) $S(t)$ is injective for $t \geq 0$

(v) There exists $B_\rho^V = \{u \mid \|u\| \leq \rho\} \subset V$ which is an absorbing set, i.e., for every $u_0 \in H$, there exists $t_0(|u_0|)$ such that, for $t \geq t(|u_0|)$, $S(t)u_0 \in B_\rho^V$.

(vi) $S(t):H \to H$ are continuous, for $t \geq 0$.

Proof. Property (i) follows from uniqueness of solutions and the fact that f is time independent. Property (iii) is a restatement of the result of Theorem 12.1. Property (iv) follows from the Backward Uniqueness Theorem (Theorem 12.2). Property (v) is a straightforward consequence of the first two energy estimates:

$$\frac{1}{2}\frac{d}{dt}|u|^2 + \nu\|u\|^2 \leq |f||u| \tag{13.5}$$

$$\frac{1}{2}\frac{d}{dt}\|u\|^2 + \nu|Au|^2 \leq |f||Au| + c|u|^{1/2}\|u\||Au|^{3/2} \tag{13.6}$$

We obtain

$$\frac{d}{dt}|u|^2 + \nu\|u\|^2 \leq \frac{|f|^2}{\nu\lambda_1} \tag{13.7}$$

From (13.7) we deduce first that

$$\nu\int_s^t \|u\|^2 ds \leq \frac{|f|^2}{\nu\lambda_1}(t-s) + |u(s)|^2 \tag{13.8}$$

and also, using $\|u\|^2 \geq \lambda_1 |u|^2$ that

(13.9) $$|u(t)|^2 \leq |u_0|^2 e^{-\nu\lambda_1 t} + \frac{|f|^2}{\nu^2\lambda_1^2}(1 - e^{-\nu\lambda_1 t}).$$

From (13.10) it follows that, for any $u_0 \in H$, if $t \geq t_0(|u_0|)$ then

(13.10) $$|u(t)|^2 \leq 2\frac{|f|^2}{\nu^2\lambda_1^2}.$$

Now we deduce from (13.6) that

(13.11) $$\frac{d}{dt}\|u\|^2 + \nu|Au|^2 \leq \frac{2|f|^2}{\nu} + \frac{c}{\nu^3}|u|^2\|u\|^4$$

If $t \geq t_0$ we obtain, from Gronwall's inequality and (13.10) an a priori bound on the growth of $\|u\|^2$ on bounded time intervals. Let us take a time length of, say $2/\nu\lambda_1$. For $t \geq t_0$ it follows from (13.8), (13.10) that

$$\int_\tau^{\tau + \frac{2}{\nu\lambda_1}} \frac{c}{\nu^3}|u|^2\|u\|^2 d\sigma \leq \frac{c}{\nu^3}\cdot\frac{2|f|^2}{\nu^2\lambda_1^2}\left(\frac{|f|^2}{\nu\lambda_1}\cdot\frac{2}{\nu^2\lambda_1} + \frac{2|f|^2}{\nu^3\lambda_1^2}\right)$$

Denoting by G the nondimensional number

(13.12) $$G = \frac{|f|}{\nu^2\lambda_1}$$

we get that, for $\tau \geq t_0(|u_0|)$

$$\int_\tau^{\tau + \frac{2}{\nu\lambda_1}} \frac{c}{\nu^3}|u|^2\|u\|^2 d\sigma \leq (8c)G^4.$$

Then from (13.11) we deduce, using Gronwall's inequality that, for $\tau \geq t_0(|u_0|)$ and $t \in [\tau, \tau + \frac{2}{\nu\lambda_1}]$ we have

(13.13) $$\|u(t)\|^2 \leq \|u(\tau)\|^2 e^{8cG^4} + 4\nu^2\lambda_1 G^2.$$

On the other hand, from (13.8) we infer that in any interval of length $1/\nu\lambda_1$, $[t,t+1/\nu\lambda_1]$ with $t \geq t_0$ we find at least some τ such that $\|u(\tau)\|^2 \leq 6\nu^2\lambda_1 G^2$.

Now let us cover $[t_0,\infty)$ by the intervals $[t_k,t_{k+1}]$ where $t_k = t_0 + \frac{k}{\nu\lambda_1}$, $k = 0,1,\ldots.$ In each such interval, we find $\tau_k \in [t_k,t_{k+1}]$ such that $\|u(\tau_k)\|^2 \leq 6\nu^2\lambda_1 G^2$.

Since the distance between successive τ_k's does not exceed $2/\nu\lambda_1$ it follows from (13.13) that, for $t \in [\tau_k,\tau_{k+1}]$

$$(13.14) \quad \|u(t)\|^2 \leq \nu^2\lambda_1 G^2(6e^{8cG^4} + 4).$$

Therefore the estimate (13.14) holds for all $t \geq \tau_0$ and in particular for all $t \geq t_0(|u_0|) + 1/\nu\lambda_1 = t_1$.

Since the condition determining t_0 was $|u_0|^2 e^{-\nu\lambda_1 t_0} \leq \frac{|f|^2}{\nu^2\lambda_1^2}$, property (v) is proven with ρ given by

$$(13.15) \quad \rho = \nu\lambda_1^{1/2} G(6e^{8cG^4} + 4)^{1/2}$$

and t_0 given by

$$(13.16) \quad \nu\lambda_1 t_0 \geq \log \frac{\nu^2\lambda_1^2|u_0|^2}{|f|^2} + 1.$$

If $f = 0$ then for any u_0, $S(t)u_0$ tend exponentially to zero in both the H and V norms.

For $u_0 \in V$, let us denote by $A(t)$ the operator

$$(13.17) \quad A(t)v = \nu Av + B(S(t)u_0,v) + B(v,S(t)u_0).$$

A proper notation would be $A(t,u_0)$. Thus (13.3) can be written as

$$(13.18) \quad \frac{dv}{dt} + A(t)v + 0.$$

Proposition 13.2. Let $u_0 \in V$, $v_0 \in H$. There exist constants k_1, k_2 depending on ν, $\|u_0\|$, $|f|$ such that the solution $v(t)$ to (13.3), (13.4) satisfies

(i) v is a real analytic $\mathcal{D}(A)$ valued function for $t > 0$

(ii) $|v(t)| \leq e^{k_1 \nu \lambda_1 t} |v_0|$ for all $t \geq 0$

(iii) $\dfrac{\|v(t)\|}{\sqrt{\lambda_1}} \leq \left(k_2 + \dfrac{1}{\nu\lambda_1 t}\right)^{1/2} e^{k_1 \nu\lambda_1 t} |v_0|$, for all $t > 0$.

Proof. We will give only a sketch of the proof because it is very similar to the proof of the corresponding result for $S(t)u_0 = u(t)$. The existence of solutions follows from linear theory. One can easily devise a Galerkin approximation, also. The estimate (ii) follows from the first energy estimate

$$(13.19) \quad \frac{d}{dt}|v|^2 + \nu\|v\|^2 \leq c\,\frac{\|u\|^2}{\nu}|v|^2.$$

Thus k_1 can be chosen to be

$$k_1 = \frac{c}{2} \sup_{t \geq 0} \frac{\|S(t)u_0\|^2}{\nu^2 \lambda_1}$$

The proof of (v) of Proposition 13.1 shows how to bound k_1 in terms of $\dfrac{\|u_0\|}{\nu\sqrt{\lambda_1}}$ and $\dfrac{|f|}{\nu^2\lambda_1}$. The estimate (iii) follows from the second energy estimate:

$$\frac{1}{2}\frac{d}{dt}\|v\|^2 + \nu|Av|^2 \leq |B(v,u,Av)| + |B(u,v,Av)|$$

$$\leq c\|u\||v|^{1/2}|Av|^{3/2} + c|u|^{1/2}\|u\|^{1/2}\|v\|^{1/2}|Av|^{3/2}$$

$$\leq c\left(\frac{\|u\|}{\lambda_1^{1/4}} + |u|^{1/2}\|u\|^{1/2}\right)\|v\|^{1/2}|Av|^{3/2}$$

$$\leq \frac{\nu}{2}|Av|^2 + \frac{c}{\nu^3}\left[\frac{\|u\|}{\lambda_1^{1/4}} + |u|^{1/2}\|u\|^{1/2}\right]^4\|v\|^2$$

The second energy inequality is

$$(13.20)\qquad \frac{d}{dt}\|v\|^2 + \nu|Av|^2 \le k_2\nu\lambda_1\|v\|^2$$

with

$$k_2 = \sup_{t>0}\left[\frac{\|u\|}{\nu\lambda_1^{1/2}} + \frac{|u|^{1/2}\|u\|^{1/2}}{\nu\lambda_1^{1/4}}\right]^4$$

From (13.9) and (ii) we obtain

$$\nu\int_{t_1}^{t_2}\|v\|^2 \le 2k_1(\nu\lambda_1)|v_0|^2\int_{t_1}^{t_2} e^{2k_1\nu\lambda_1 t}\,dt + |v(0)|^2 e^{2k_1\nu\lambda_1 t_1}.$$

Thus

$$(13.21)\qquad \nu\int_{t_1}^{t_2}\|v\|^2 \le |v_0|^2 e^{2k_1\nu\lambda_1 t_2}.$$

Integrating (13.20) we get

$$\|v(t_2)\|^2 - \|v(t_1)\|^2 \le k_2\lambda_1\nu\int_{t_1}^{t_2}\|v\|^2$$

Using (13.21)

$$\|v(t_2)\|^2 \le \|v(t_1)\|^2 + k_2\lambda_1|v_0|^2 e^{2k_1\nu\lambda_1 t_2}.$$

Integrating $\int_0^{t_2} dt_1$ we obtain

$$t_2\|v(t_2)\|^2 \le \frac{|v_0|^2}{\nu} e^{2k_1\nu\lambda_1 t_2} + k_2\lambda_1 t_2|v_0|^2 e^{2k_1\nu\lambda_1 t_2}.$$

Multiplying by ν:

$$\nu\lambda_1 t_2\frac{\|v(t_2)\|^2}{\lambda_1} \le |v_0|^2(1 + k_2\nu\lambda_1 t_2)e^{2k_1\nu\lambda_1 t_2}$$

which is (iii).

Let now ϕ be a smooth function defined on an open set $D \subset \mathbb{R}^N$, $N \ge 1$ and taking values in V. Let Σ_0 be the image of ϕ. Let $\Sigma_t = S(t)\Sigma_0$. Let us denote by $(\cdot;\cdot)$ and $|\cdot|$ the scalar product and norm in $\wedge^N H$ (the N-th exterior product of H). The volume element in Σ_t is

$$\left|\frac{\partial}{\partial\alpha_1}(S(t)\phi(\alpha))\wedge\frac{\partial}{\partial\alpha_2}(S(t)\phi(\alpha))\wedge\cdots\wedge\frac{\partial}{\partial\alpha_N}(S(t)\phi(\alpha))\right|d_N\alpha\ ,$$

where $d_N\alpha = d\alpha_1 \cdots d\alpha_N$ is the volume element in $\mathbb{R}^N$. The functions

$$v_i(t) = \frac{\partial}{\partial\alpha_i}(S(t)\phi(\alpha)),\ i = 1,2,\ldots,N$$

satisfy the linearized equation (13.3) along $u(t) = S(t)\phi(\alpha)$. Therefore in order to study the time evolution of the volume element of an N-dimensional surface transported by S(t) we are lead to study the time evolution of

$$|v_1(t)\wedge\cdots\wedge v_N(t)|$$

where $v_1,\ldots,v_N$ satisfy equation (13.3) along some $u(t) = S(t)u_0$.

Let us recall the formula

$$(13.22)\qquad (v_1\wedge\cdots\wedge v_N; w_1\wedge\cdots\wedge w_N) = \det(v_i,w_j)_{\substack{i=1,\ldots,N\\ j=1,\ldots,N}}.$$

If $v_1,\ldots,v_N$ are elements of H we denote by $Q(v_1,\ldots,v_N)$ the orthogonal projector in H onto the linear space spanned by the vectors $v_1,\ldots,v_N$. If $T:\mathcal{D}(T)\subset H\to H$ is an operator in H then we define an operator in $\wedge^N H$ by

$$T_N = T\wedge I\wedge I\wedge\cdots\wedge I + I\wedge T\wedge I\wedge\cdots + I\wedge I\wedge\cdots\wedge T.$$

T_N will act on a monome $v_1\wedge\cdots\wedge v_N$ as

$$T_N(v_1\wedge\cdots\wedge v_N) = Tv_1\wedge v_2\wedge\cdots\wedge v_N + \cdots + v_1\wedge\cdots\wedge Tv_N\ .$$

The following formula is proven in [CF1]

Lemma 13.3. let $v_1,\ldots,v_N$ be elements of $\mathcal{D}(T)$. Then

$$(13.23)\quad (T_N(v_1\wedge\cdots\wedge v_N); v_1\wedge\cdots\wedge v_N) = |v_1\wedge\cdots\wedge v_N|^2\ \mathrm{Tr}(TQ(v_1,\ldots,v_N))$$

As a consequence we can write the equation satisfied by the Wronskian $|v_1(t)\wedge \cdots \wedge v_N(t)|$.

Lemma 13.4. Let $v_1(t),\ldots,v_N(t)$ be solutions of

$$\begin{cases} \dfrac{dv_i}{dt} + A(t)v_i = 0 \\ v_i(0) = v_i^0 \end{cases} \qquad i = 1,\ldots,N$$

where $A(t)v = \nu Av + B(S(t)u_0,v) + B(v,S(t)u_0)$ and $u_0 \in V$. Then the Wronskian $|v_1(t)\wedge \cdots \wedge v_N(t)|$ is either identically zero or never vanishes and satisfies

$$(13.24) \quad \frac{1}{2}\frac{d}{dt}|v_1\wedge \cdots \wedge v_N|^2 + |v_1\wedge \cdots \wedge v_N|^2 \, \mathrm{Tr}(A(t)Q(v_1,\ldots,v_N)) = 0$$

Proof. The equation satisfied by $v_1\wedge \cdots \wedge v_N$ is

$$\frac{d}{dt}(v_1\wedge \cdots \wedge v_N) + A(t)_N (v_1\wedge \cdots \wedge v_N) = 0$$

Thus (13.14) is a consequence of (13.23).

Denoting the lower order term in $A(t)$ by

$$(13.25) \quad L(t)v = B(S(t)u_0,v) + B(v,S(t)u_0)$$

we have

$$\mathrm{Tr}(A(t)Q(v_1,\ldots,v_N)) = \nu\mathrm{Tr}(AQ(v_1,\ldots,v_N)) + \mathrm{Tr}(L(t)Q(v_1,\ldots,v_N)).$$

Now $\mathrm{Tr}\, AQ(v_1,\ldots,v_N)$ can be computed, using again (13.23)

$$(13.26) \quad \mathrm{Tr}\, AQ(v_1,\ldots,v_N) = \frac{(A_N(v_1\wedge \cdots \wedge v_N);v_1\wedge \cdots \wedge v_N)}{|v_1\wedge \cdots \wedge v_N|^2}$$

Now the first eigenvalue of A_N is $\lambda_1 + \lambda_2 + \cdots + \lambda_N$. Therefore

$$(13.27)\quad \operatorname{Tr} AQ(v_1,\ldots,v_N) \geq \lambda_1 + \cdots + \lambda_N.$$

Let us remark here that we could have derived entirely similar results in V instead of H. The volume element would have to be computed in $\Lambda^N V$. The orthogonal projection $Q(v_1,\ldots,v_N)$ will have to be orthogonal with respect to the scalar product in V. The first eigenvalue of A_N in $\Lambda^N V$ is still $\lambda_1 + \cdots + \lambda_N$.

From Theorem 4.11, we know that $\lambda_j \geq c_0 j \lambda_1$ (we are in the two-dimensional case). Thus

$$(13.28)\quad \lambda_1 + \cdots + \lambda_N \geq c_0\lambda_1 \frac{N(N+1)}{2}.$$

We will give lower bounds for

$$\frac{1}{t}\int_0^t \operatorname{Tr}(A(s)Q(v_1(s),\ldots,v_N(s))ds.$$

From (13.24) it follows that

$$(13.29)\quad |v_1\wedge\cdots\wedge v_N|^2(t) \leq |v_1^0\wedge\cdots\wedge v_N^0|^2\exp(-t\gamma_N(t))$$

where $\gamma_N(t)$ is a function depending actually on N, t, u_0, $v_1^0,\ldots,v_N^0$:

$$(13.30)\quad \gamma_N(t) = \frac{1}{t}\int_0^t \operatorname{Tr}(A(s)Q(v_1(s),\ldots,v_N(s))ds.$$

It is clear that a lower bound of the type

$$(13.31)\quad \gamma_N(t) \geq \nu\lambda_1 c_N$$

for all $t \geq t_0 = t_0(\|u_0\|)$, and $c_N > 0$ will imply exponential decay for volume elements.

We will present two ways of estimating the averages

$$\frac{1}{t}\int_0^t \operatorname{Tr}(A(s)Q(v_1(s),\ldots,v_N(s))ds.$$

Let us first observe that, since $B(S(t)u_0, \cdot)$ is an antisymmetric operator in H the trace

$$Tr(L(s)Q(v_1(s),\dots,v_N(s))$$

is equal to

$$Tr(B(\cdot\ ,S(t)u_0)Q(v_1(s),\dots,v_N(s))).$$

Let us fix $s > 0$ and omit the dependence on s of various quantities, for the moment.

$$Tr\ L(s)Q(v_1(s),\dots,v_N(s)) = Tr\ B(\cdot,S(s)u_0)Q(v_1(s),\dots,v_N(s))$$

$$= \sum_{i=1}^{N} B(\varphi_i, u, \varphi_i).$$

Here $\varphi_1,\dots \varphi_N$ are an orthonormal family of functions $\varphi_i \in V$, $(\varphi_i, \varphi_j) = \delta_{ij}$ spanning the linear span of $v_1(s),\dots,v_N(s)$. The element $u = S(s)u_0$. A direct estimate would give

$$|\sum_{i=1}^{N} B(\varphi_i,u,\varphi_i)| \le \sum_{i=1}^{N} |B(\varphi_i,u,\varphi_i|$$

$$\le c \sum_{i=1}^{N} |\varphi_i|\|\varphi_i\|\|u\| = c \sum_{i=1}^{N} \|\varphi_i\|\|u\|$$

$$\le c(\sum_{i=1}^{N} \|\varphi_i\|^2)^{1/2} N^{1/2}\|u\|.$$

On the other hand

$$Tr\ AQ(v_1(s),\dots,v_N(s)) = \sum_{i=1}^{N} (A\varphi_i,\varphi_i) = \sum_{i=1}^{N} \|\varphi_i\|^2.$$

Therefore,

$$(13.32) \quad |Tr\ L(s)Q(v_1(s),\dots,v_N(s))|$$

$$\le cN^{1/2}\|S(s)u_0\|\cdot(Tr\ AQ(v_1,\dots,v_N(s)))^{1/2}$$

From (13.22) it follows that

$$\frac{1}{t}\int_0^t |\mathrm{Tr}\, L(s)Q(v_1(s),\ldots,v_N(s))|ds$$

$$\leq cN^{1/2}\Big(\frac{1}{t}\int_0^t \|S(s)u_0\|^2 ds\Big)^{1/2}\Big(\frac{1}{t}\int_0^t \mathrm{Tr}\, AQ(v_1(s),\ldots,v_N(s))ds\Big)^{1/2}.$$

Therefore

$$\frac{1}{t}\int_0^t \mathrm{Tr}\, A(s)Q(v_1(s),\ldots,v_N(s)) = \frac{1}{t}\int_0^t \mathrm{Tr}\, \nu AQ(v_1(s),\ldots,v_N(s))ds$$

$$+\frac{1}{t}\int_0^t \mathrm{Tr}\, L(s)Q(v_1(s),\ldots,v_N(s))ds$$

$$\geq \frac{\nu}{2}\frac{1}{t}\int_0^t \mathrm{Tr}(AQ(v_1(s),\ldots,v_N(s)) - \frac{c^2 N}{2\nu}\frac{1}{t}\int_0^t \|S(s)u_0\|^2 ds$$

$$\geq \frac{\nu}{2}(\lambda_1 + \cdots + \lambda_N) - \frac{c^2}{2\nu} N \frac{1}{t}\int_0^t \|S(s)u_0\|^2 ds$$

$$\geq \frac{c_0\nu\lambda_1}{4} N(N+1) - \frac{c^2}{2\nu} N \frac{1}{t}\int_0^t \|S(s)u_0\|^2 ds.$$

Now, from the first energy inequality (13.7)

$$\frac{1}{t}\int_0^t \|S(s)u_0\|^2 ds \leq \frac{|f|^2}{\nu^2\lambda_1} + \frac{|u_0|^2}{\nu t}.$$

We obtain

$$\frac{1}{t}\int_0^t \mathrm{Tr}(A(s)Q(v_1(s),\ldots,v_N(s))ds$$

$$\geq \frac{c_0\nu\lambda_1}{4} N(N+1) - \frac{c^2}{2}\nu\lambda_1 N\Big[\frac{|f|^2}{\nu^4\lambda_1^2} + \frac{|u_0|^2}{\nu^3\lambda_1 t}\Big]$$

$$= N\nu\lambda_1\Big\{\frac{c_0}{4}(N+1) - \frac{c^2}{2}\Big[G^2 + \frac{|u_0|^2}{\nu^2(\nu\lambda_1 t)}\Big]\Big\}.$$

We take $t_0 = t_0(|u_0|)$ to be defined by

$$\nu\lambda_1 t_0 = \frac{|u_0|^2}{\nu^2}\cdot\frac{1}{G^2}. \tag{13.33}$$

We proved

$$(13.34)\quad \frac{1}{t}\int_0^t \mathrm{Tr}(A(s)Q(v_1(s),\ldots,v_N(s))ds \geq c_2 N\nu\lambda_1(N+1 - c_3G^2)$$

for all $t \geq t_0(|u_0|)$. The constants are $c_2 = c_0/4$, $c_3 = 4c^2/c_0$, with c_0 the constant occurring in Theorem 4.11 and c the constant in the estimate $|B(\varphi,u,\varphi)| \leq c|\varphi|\|\varphi\|\|u\|$. The estimate (13.34) implies that,

(13.35) If $N \geq c_3G^2$, then $\gamma_N(t) \geq c_2N\nu\lambda_1$, for $t \geq t_0(|u_0|)$.

This would mean that the exponential decay of volume elements would start at dimensions $\geq c_3G^2$.

Actually, replacing the Sobolev estimates at one step in the preceding argument by a Lieb-Thirring inequality yields an improvement. The Lieb-Thirring inequality we refer to is the following [L-T].

Theorem 13.5 (A Lieb-Thirring Inequality). Let $\varphi_1,\ldots,\varphi_N$ be $H^1(\mathbb{R}^2)$ functions. Assume that the φ_i's are orthonormal in L^2:

$$\int \varphi_i(x)\ \varphi_j(x)dx = \delta_{ij}.$$

Then there exists a constant c_L, independent of N, such that

$$(13.36)\quad \int \Big(\sum_{i=1}^{N}|\varphi_i(x)|^2\Big)^2dx \leq c_L \sum_{i=1}^{N}\int|\nabla\varphi_i|^2dx.$$

We will use (13.36) with $\varphi_i \in V$ which implies $\varphi_i \in H_0^1(\Omega)^2$, $\Omega \subset \mathbb{R}^2$ bounded. We return to the estimate of $\gamma_N(t)$. We compute

$$\mathrm{Tr}\ L(s)Q(v_1(s),\ldots,v_N(s)) = \sum_{i=1}^{N} B(\varphi_i,u,\varphi_i) = \sum_{i=1}^{N}\int_\Omega \langle\nabla u(x)\ \varphi_i,\ \varphi_i\rangle$$

Therefore

$$|\mathrm{Tr}\ L(s)Q(v_1(s),\ldots,v_N(s))| \leq \int_\Omega |\nabla u(x)|\left(\sum_{i=1}^N |\varphi_i(x)|^2\right)dx$$

$$\leq \left[\int_\Omega \left(\sum_{i=1}^N |\varphi_i(x)|^2\right)^2\right]^{1/2} \|u\|$$

$$\leq \text{with Lieb-Thirring} \quad c_L^{1/2}\left(\sum_{i=1}^N \|\varphi_i\|^2\right)^{1/2} \|u\|$$

$$= c_L^{1/2}(\mathrm{Tr}\ AQ(v_1(s),\ldots,v_N(s)))^{1/2} \|S(s)u_0\|.$$

Thus the Lieb-Thirring inequality (13.36) allows us to replace (13.32) by

$$(13.37) \quad |\mathrm{Tr}\ L(s)Q(v_1(s),\ldots,v_N(s))| \leq c_L^{1/2} \|S(s)u_0\|(\mathrm{Tr}\ AQ(v_1(s),\ldots,v_N(s)))^{1/2}$$

Notice that the difference between (13.32) and (13.37) is in the absence of $N^{1/2}$ in (13.37). (Aslo the constant is now the one appearing in (13.36).) We can repeat exactly the steps which followed (13.32) using (13.37). We keep the same definition (13.33) of t_0 and get that

$$(13.38) \quad \gamma_N(t) \geq \nu\lambda_1\left(\frac{c_0}{4} N(N+1) - c_L G^2\right) \quad \text{provided } t \geq t_0(|u_0|).$$

We proved

Theorem 13.6. Let $\Omega \subset \mathbb{R}^2$ be bounded, open, of class C^2. Let $u_0 \in V$. Let $S(t)u_0 = u(t)$ be a solution of the Navier-Stokes equations

$$\begin{cases} \dfrac{du}{dt} + \nu Au + B(u,u) = f \\ \\ u(0) = u_0 \end{cases}$$

with $f \in H$. Denote by G

$$G = \frac{|f|}{\nu^2\lambda_1}.$$

Let N be a positive integer satisfying

(13.39) $N \geq c_4 G$

with c_4 an absolute constant ($c_4 = 2\sqrt{\frac{c_L}{c_0}}$). Let $v_1,\ldots,v_N$ be arbitrary solutions of the linearized equation

$$\frac{dv}{dt} + \nu Av + B(u,v) + B(v,u) = 0$$

with initial data

$$v_i(0) = v_i^0 \in H \qquad i = 1,\ldots,N.$$

Then the volume elements $|v_1(t)\wedge \cdots \wedge v_N(t)|$ decay exponentially. More precisely

(13.40) $|v_1(t)\wedge \cdots \wedge v_N(t)|^2 \leq |v_1^0\wedge \cdots \wedge v_N^0|^2 \exp(-\frac{c_0}{4}\nu\lambda_1 Nt)$

for all $t \geq t_0(|u_0|)$ defined by

(13.41) $t_0(|u_0|) = \frac{|u_0|^2}{\nu^3\lambda_1}\frac{1}{G^2}$.

If $G = 0$ the exponential decay starts at $N = 1$ and t_0 may be chosen to be $\frac{|u_0|^2}{\nu^3\lambda_1}\cdot\frac{c_0}{2c_L}$.

Now we will present an improvement of the result in Theorem 13.6 in the case of periodic boundary conditions. The result will be that exponential decay of volume elements starts at $N \sim G^{2/3}(\log G)^{1/3}$. the idea of the proof is to compute everything in V instead of H and use the Lieb-Thirring inequality used before together with another inequality [C2] which is an L^∞ generalization of some L^p inequalities of Lieb [L1].

As we mentioned earlier we can compute volume elements in V. We consider the equation

(13.42) $\frac{dv}{dt} + \nu Av + B(u,v) + B(v,u) = 0$

along the solution u(t) of

$$(13.43)\quad \begin{cases} \frac{du}{dt} + \nu Au + B(u,u) = f \\ u(0) = u_0 \end{cases}$$

in the two-dimensional periodic case. Let N be a positive integer and let $v_i^0 \in V$ for $i = 1,\ldots,N$. We recall the notation $((v,w))$ for the scalar product in V

$$((v,w)) = (A^{1/2}v, A^{1/2}w).$$

We consider the volume elements

$$\|v_1 \wedge \ldots \wedge v_N\|^2 = \det((v_i,v_j))_{i,j = 1,\ldots,N}$$

computed for $v_i(t) = v_i$ solutions of (13.42). Then, as in (13.24) we have

$$(13.44)\quad \frac{1}{2}\frac{d}{dt}\|v_1 \wedge \ldots \wedge v_N\|^2 + \|v_1 \wedge \ldots \wedge v_N\|^2 \operatorname{Tr}(A(t)Q^V(v_1,\ldots,v_N)) = 0$$

where $Q^V(v_1,\ldots,v_N)$ is the orthogonal projection in V on the linear space spanned by the vectors $v_1,\ldots,v_N$. The operator A(t) is still given by the expression

$$(13.45)\quad A(t)v = \nu Av + B(S(t)u_0,v) + B(v,S(t)u_0)$$

but it is viewed as an operator in V. We need to give a lower bound for

$$\frac{1}{t}\int_0^t \operatorname{Tr}(A(s)Q^V(v_1(s),\ldots,v_N(s)))ds.$$

Let us note first that, in the periodic case the identity

$$(13.46)\quad (B(v,v),Av) = 0$$

holds for all $v \in \mathcal{D}(A)$. The proof is done by integration by parts.

Differentiating (13.46) we obtain

$$(13.47)\quad (B(u,v),Av) + (B(v,u),Av) + (B(v,v),Au) = 0.$$

Let $\varphi_1,\ldots,\varphi_N$ be a family of orthonormal elements of V

$$(13.48)\quad ((\varphi_i,\varphi_j)) = \delta_{ij}$$

such that their linear span coincides with the span of the vectors $v_1(s),\ldots,v_N(s)$. In order to compute $\mathrm{Tr}\ L(s)Q^V(v_1(s),\ldots,v_N(s))$ we use (13.48):

$$\mathrm{Tr}\ L(s)Q^V(v_1(s),\ldots,v_N(s)) = \sum_{i=1}^{N}(B(u,\varphi_i) + B(\varphi_i,u),A\varphi_i)$$
$$= -\sum_{i=1}^{N} B(\varphi_i,\varphi_i,Au).$$

Now, since we are in the periodic case

$$-\sum_{i=1}^{N} B(\varphi_i,\varphi_i,Au) = \sum_{i=1}^{N}\int(\varphi_i,\nabla\varphi_i)\Delta u$$

Let us denote by $\rho(x)$ the function

$$(13.49)\quad \rho(x) = \sum_{i=1}^{N}|\varphi_i(x)|^2$$

and by $\sigma(x)$ the function

$$(13.50)\quad \sigma(x) = \sum_{i=1}^{N}|\nabla\varphi_i(x)|^2.$$

We have

$$(13.51)\quad |\mathrm{Tr}\ L(s)Q^V(v_1(s),\ldots,v_N(s))| \le |\rho|_{L^\infty}^{1/2}|\sigma|_{L^2}^{1/2}|\Delta u|_{L^{4/3}}\ .$$

Notice that, since φ_i are orthonormal in V

$$\int (\nabla \varphi_i)(\nabla \varphi_j) = \delta_{ij}$$

and therefore from the Lieb-Thirring inequality (13.36) for $(\nabla \varphi_i)$ it follows that

$$(13.52) \quad |\sigma|_{L^2}^{1/2} \leq c\left(\sum_{i=1}^{N} |A\varphi_i|^2\right)^{1/4} = c(\mathrm{Tr}\ AQ^V(v_1(s),\dots,v_N(s)))^{1/4}.$$

On the other hand, in order to estimate $|\rho|_{L^\infty}$ we use an inequality ([C2]) of the following type:

Lemma 13.7. Let $\{\varphi_i\}_{i=1,\dots,N}$ be a sequence of functions belonging to $\mathcal{D}(A)$ and orthonormal in V. Let ρ be defined by $\rho(x) = \sum_{i=1}^{N} |\varphi_i(x)|^2$. There exists a constant, independent of N such that

$$(13.53) \quad |\rho|_{L^\infty} \leq c\left(1 + \log \frac{1}{\lambda_1} \sum_{i=1}^{N} |A\varphi_i|^2\right).$$

The proof of this lemma is rather technical and will not be given here. We estimate the term $|\Delta u|_{L^{4/3}}$ by

$$(13.54) \quad |\Delta u|_{L^{4/3}} \leq c\lambda_1^{-1/4}|Au|$$

Now, combining (13.51), (13.52), (13.53), and (13.54) we get

$$(13.55) \quad |\mathrm{Tr}\ L(s)Q(s)|$$

$$\leq c_5|AS(s)u_0|\left(1 + \log \frac{1}{\lambda_1} \mathrm{Tr}\ AQ(s)\right)^{1/2}\left(\frac{1}{\lambda_1} \mathrm{Tr}\ AQ(s)\right)^{1/4}.$$

This is the analogue of the estimates (13.32) and (13.37). We denoted, for simplicity $Q(s) = Q^V(v_1(s),\dots,v_N(s))$. Let us denote, in order to alleviate the computation by

(13.56) $x(s) = \frac{1}{\lambda_1} \operatorname{Tr} (AQ^V(v_1(s),\ldots,v_N(s)))$.

Note that, in view of (13.28)

(13.57) $x(s) \geq \frac{c_0}{2} (N+1)N$.

Let us assume that N is chosen such that

(13.58) $\frac{c_0}{2} N(N+1) \geq 1$.

In the following string of inequalties we will use, at some point, the concavity of the function $g(x) = x^{1/2}(1 + \log x)$ for $x \geq 1 > 1/e$ (the function $g(x)$ is concave for $x > 1/e$) and Jensen's inequality

$$\frac{1}{t}\int_0^t g(x(s))ds \leq g(\frac{1}{t}\int_0^t x(s)ds)$$

which can be applied since $x(s) \geq 1$ on $[0,t]$. The inequalities are as follows:

$$\frac{1}{t}\int_0^t \operatorname{Tr} AQ^V(v_1(s),\ldots,v_N(s))ds$$

$$= \frac{\nu}{t}\int_0^t \operatorname{Tr} AQ(s)ds + \frac{1}{t}\int_0^t \operatorname{Tr} L(s)Q(s) \geq \text{(with (13.55))}$$

$$\geq \frac{\nu\lambda_1}{t}\int_0^t x(s)ds - c_5\frac{1}{t}\int_0^t (1 + \log x(s))^{1/2}(x(s))^{1/4}|A(S(s)u_0)|ds$$

$$\geq \frac{\nu\lambda_1}{t}\int_0^t x(s)ds - c_5(\frac{1}{t}\int_0^t |AS(s)u_0|^2ds)^{1/2}(\frac{1}{t}\int_0^t g(x(s))ds)^{1/2}$$

$\geq$ (with Jensen's inequality)

$$\geq \frac{\nu\lambda_1}{t}\int_0^t x(s)ds - c_5(\frac{1}{t}\int_0^t |A(S(s)u_0)|^2ds)^{1/2}(g(\frac{1}{t}\int_0^t x(s)ds)^{1/2}).$$

Let us denote

$$(13.59)\qquad c_5\left(\frac{1}{t}\int_0^t |AS(s)u_0|^2 ds\right)^{1/2} = \alpha(t,u_0) = \alpha$$

and

$$(13.60)\qquad \frac{1}{t}\int_0^t x(s)ds = m(t) = m.$$

Then we have

$$\frac{1}{t}\int_0^t \operatorname{Tr} AQ^V(v_1(s),\dots,v_N(s))ds \ge \nu\lambda_1 m^{1/4}(1+\log m)^{1/2}\left(\frac{m^{3/4}}{(1+\log m)^{1/2}} - \frac{\alpha}{\nu\lambda_1}\right)$$

and therefore, since $m \ge \frac{c_0}{2} N(N+1) \ge 1$, we will have

$$(13.61)\qquad \frac{1}{t}\int_0^t \operatorname{Tr} AQ^V(v_1(s),\dots,v_N(s))ds \ge \nu\lambda_1\left(\frac{c_0}{2} N(N+1)\right)^{1/4} > 0$$

provided

$$\frac{m^{3/4}}{(1 + \log m)^{1/2}} \ge \frac{\alpha}{\nu\lambda_1} + 1.$$

Now the function $h(m) = \dfrac{m^{3/4}}{(1 + \log m)^{1/2}}$ satisfies $h'(m) > 0$ for $m \ge 1$ and $h(1) = 1$. Therefore, $h(m) \ge \frac{\alpha}{\nu\lambda_1} + 1$ if $m \ge m_0$ where m_0 is the solution of $h(m_0) = \frac{\alpha}{\nu\lambda_1} + 1$. Now if $\dfrac{m_0^{3/4}}{(1 + \log m_0)^{1/2}} = \frac{\alpha}{\nu\lambda_1} + 1$ then $m_0 = \left(\frac{\alpha}{\nu\lambda_1} + 1\right)^{4/3}(1 + \log m_0)^{2/3}$. Thus

$$\log m_0 = \frac{4}{3}\log\left(\frac{\alpha}{\nu\lambda_1} + 1\right) + \frac{2}{3}\log(1 + \log m_0) \le \frac{4}{3}\log\left(\frac{\alpha}{\nu\lambda_1} + 1\right) + \frac{2}{3}\log m_0.$$

It follows that $\log m_0 \le 4 \log\left(\frac{\alpha}{\nu\lambda_1} + 1\right)$ and consequently that

$$(13.62)\qquad m_0 \le \left(\frac{\alpha}{\nu\lambda_1} + 1\right)^{4/3}\left(1 + 4\log\left(\frac{\alpha}{\nu\lambda_1} + 1\right)\right)^{2/3}.$$

Let us denote by

$$(13.63)\qquad \beta(u_0) = \frac{1}{\nu\lambda_1} \limsup_{t \to \infty} \left(\frac{1}{t}\int_0^t |AS(s)u_0|^2 ds\right)^{1/2}.$$

Note that $\beta = \frac{1}{c_5\nu\lambda_1} \limsup_{t \to \infty} \alpha(t,u_0)$. The preceding argument shows that the inequality (13.61) is valid provided

$$(13.64)\qquad \frac{c_0}{2} N(N+1) \geq 1 + (c_5\beta + 1)^{4/3}(1 + 4 \log(c_5\beta + 1))^{2/3}$$

In order of magnitude terms, (16.64) implies that exponential decay of volume elements (in V) along a trajectory $S(t)u_0$ takes place if the dimension N of the volume element is at least of the order of $\beta^{2/3}(\log \beta)^{1/3}$ when β is the time average of $|AS(t)u_0|$ defined in (13.63).

We can give easily an upper bound for β. Taking the second energy equation for $u(t) = S(t)u_0$, that is taking the scalar product of (13.43) with $Au(t)$ and using (13.64) we obtain

$$(13.65)\qquad \nu \int_0^t |Au(s)|^2 ds \leq \|u(0)\|^2 + t\,\frac{|f|^2}{\nu}.$$

Dividing by t and taking $\limsup_{t \to \infty}$ we infer

$$(13.66)\qquad \beta(u_0) \leq G.$$

Note that $G = \frac{|f|}{\nu^2\lambda_1}$ is independent of u_0. Also note that (13.65) implies that the expression $\alpha(t)$ can be bounded from above by $(c_5\nu\lambda_1)^{-1}\beta(u_0) + \delta$, for any $\delta > 0$, provided $t \geq t_1(\delta,\|u_0\|)$; thus uniformly for u_0 in bounded sets of V. We proved

Theorem 13.7. Let $u(t) = S(t)u_0$ be a solution of the two-dimensional Navier-Stokes equation with periodic boundary conditions

$$\begin{cases} \frac{du}{dt} + \nu Au + B(u,u) = f \\ u(0) = 0 \end{cases}$$

Assume $f \in H$ and let $G = \frac{|f|}{\nu^2 \lambda_1}$. Assume that the positive integer N satisfies

(16.64) $N \geq c_6(1 + G)^{2/3}(\log(G+2))^{1/3}$

with an appropriate absolute constant c_6. Let $v_1(t),\ldots,v_N(t)$ be solutions of the linearized equation along $S(t)u_0 = u(t)$

$$\frac{dv}{dt} + \nu Av + B(u,v) + B(v,u) = 0$$

with initial data $v_i(0) = v_i^0$ belonging to V, $i = 1,..,N$. Then the volume elements $\|v_1(t) \wedge \ldots \wedge v_N(t)\|$ decay exponentially. More precisely

(13.68) $\| v_1(t) \wedge \ldots \wedge v_N(t) \|^2 \leq \| v_1^0 \wedge \ldots \wedge v_N^0 \|^2 \exp(-\nu\lambda_1(c_0 N(N+1))^{1/4} t)$

provided $t \geq t_1(\|u_0\|)$.

We conclude this section with a brief description of the three-dimensional Navier-Stokes case. Let us take a strong solution $u(t) = S(t)u_0$ of the Navier-Stokes equation in $\Omega \subset \mathbb{R}^3$. Let us work in H. Taking the linearized equation and computing the volume elements $|v_1(t) \wedge \ldots \wedge v_N(t)|$ we arrive at the same equation as in the two-dimensional case ((13.24)). In order to estimate $Tr(L(s)Q(v_1(s),\ldots,v_N(s)))$ we proceed as in the two-dimnensional case

$$Tr(L(s)Q(s)) = \sum_{i=1}^{N} B(\varphi_i, u, \varphi_i)$$

with $(\varphi_i \varphi_j) = \delta_{ij}$, $\{\varphi_1,\ldots,\varphi_N\}$ spanning the same linear subspace of H as $\{v_1(s),\ldots,v_N(s)\}$. Then we use (6.20) with $s_1 = 3/4 = s_3$, $s_2 = 0$

$$|\sum_{i=1}^{N} B(\varphi_i, u, \varphi_i)| \le \sum_{i=1}^{N} c|\varphi_i|^{1/2} \|\varphi_i\|^{3/2}\|u\|$$

$$= c \sum_{i=1}^{N} \|\varphi_i\|^{3/2}\|u\| \le c\|u\|N^{1/4} (\sum_{i=1}^{N} \|\varphi_i\|^2)^{3/4}.$$

Using Young's inequality, it follows that

$$\frac{1}{t}\int_0^t \mathrm{Tr}\, A(s)Q(v_1(s),\ldots,v_N(s))ds$$

$$= \frac{\nu}{t}\int_0^t \mathrm{Tr}\, AQ(s)ds + \frac{1}{t}\int_0^t \mathrm{Tr}\, L(s)Q(s)ds$$

$$\ge \frac{\nu}{t}\int_0^t \mathrm{Tr}\, AQ(s)ds - \frac{\nu}{2t}\int_0^t \mathrm{Tr}\, AQ(s)ds - c_7 \frac{N}{\nu^3}\frac{1}{t}\int_0^t \|u(s)\|^4 ds$$

$$= \frac{\nu}{2t}\int_0^t \mathrm{Tr}\, AQ(s)ds - c_7 \frac{N}{\nu^3}\frac{1}{t}\int_0^t \|u(s)\|^4 ds.$$

Now $\mathrm{Tr}\, AQ(s) \ge \lambda_1 + \cdots + \lambda_N$ and from Theorem 4.11,

$$\lambda_1 + \cdots + \lambda_N \ge c_0\lambda_1(1^{2/3} + 2^{2/3} + \cdots N^{2/3})$$

so

$$\lambda_1 + \cdots \lambda_N \ge \frac{3}{5} c_0\lambda_1 N^{5/3}.$$

Therefore

$$\frac{1}{t}\int_0^t \mathrm{Tr}\, A(s)Q(v_1(s),\ldots,v_N(s))ds \tag{13.69}$$

$$\ge \nu\lambda_1 N(\frac{3}{10} c_0 N^{2/3} - c_7 \frac{1}{\lambda^4\lambda_1}\frac{1}{t}\int_0^t \|u(s)\|^4 ds)$$

Clearly, if

$$\frac{3}{10} c_0 N^{2/3} - c_7 \frac{1}{\nu^4\lambda_1}\frac{1}{t}\int_0^t \|u(s)\|^4 ds \ge 1 \tag{13.70}$$

then

$$(13.71)\qquad \frac{1}{t}\int_0^t \mathrm{Tr}\, A(s)Q(v_1(s),\dots,v_N(s))ds \geq \nu\lambda_1 N.$$

Now, for an arbitrary $u_0 \in V$ (or even smoother) we have, at this moment, no way of giving an a priori bound on $\frac{1}{t}\int_0^t \|S(s)u_0\|^4 ds$. Note that this is the very same quantity that controls global existence of strong solutions and uniqueness of weak solutions. Thus, decay of volume elements can be obtained without other requirements that the ones needed for global regularity.

Theorem 13.8. Let $u(t) = S(t)u_0$ be a solution to the three-dimensional Navier-Stokes equations

$$\begin{cases} \frac{du}{dt} + \nu Au + B(u,u) = f \\ u(0) = 0 \end{cases}$$

Let us assume that the quantity $\eta(u_0)$

$$(13.72)\qquad \eta(u_0) = \limsup_{t\to\infty} \frac{1}{\nu^4\lambda_1}\frac{1}{t}\int_0^t \|S(s)u_0\|^4 ds$$

is finite. Let N be a positive integer satisfying

$$(13.73)\qquad N \geq c_8(1+\eta(u_0))^{3/2}$$

with c_8 an appropriate absolute constant. Let $v_1,\dots,v_N$ be solutions of the linearized equation along $S(t)u_0$:

$$\frac{dv}{dt} + \nu Av + B(u,v) + B(v,u) = 0$$

with initial data $v_i(0) = v_i^0 \in H$, $i = 1,\dots,N$. Then the volume elements $|v_1(t)\wedge\dots\wedge v_N(t)|$ decay exponentially. More precisely,

$$(13.74)\qquad |v_1(t)\wedge\dots\wedge v_N(t)|^2 \leq |v_1^0\wedge\dots\wedge v_N^0|^2 \exp(-\nu\lambda_1 Nt)$$

for $t \geq t_2(u_0)$.

14

GLOBAL LYAPUNOV EXPONENTS. HAUSDORFF AND FRACTAL DIMENSION OF THE UNIVERSAL ATTRACTOR

The objects of study of this section are bounded invariant ($S(t)Z = Z$) sets $Z \subset H$. We will prove that if Z is bounded and invariant then Z has finite fractal and a fortiori Hausdorff dimensions. We consider first the case of two dimensional Navier-Stokes equations in order to fix ideas. In view of the property (v) of the solution map $S(t)$ (Proposition 13.1) there exists $\rho > 0$ and a ball in V such that

$$S(t)u_0 \in B_\rho^V = \{u \in V \mid \|u\| \leq \rho\}$$

for all $u_0 \in H$ and all $t \geq t_0(|u_0|)$. Let us observe that B_ρ^V is bounded (actually compact) in H and therefore there exists $T > 0$ such that $S(t)B_\rho^V \subset B_\rho^V$ for all $t \geq T$. Let us consider the set

$$X = \bigcap_{s>0} S(s)B_\rho^V. \tag{14.1}$$

Let $t > 0$. We claim $S(t)X = X$. Indeed, if $x \in X$ then $S(t)x = S(t)S(\sigma)y$ with $y = y_\sigma \in B_\rho^V$. From the semigroup property of S, $S(t)x = S(t+\sigma)y$. Thus $S(t)x \in \bigcap_{s \geq t} S(s)B_\rho^V$. Now if $s \leq t$ then $S(t)x = S(s)S(t-s)x$ and it is enough to check that $S(t-s)x \in B_\rho^V$. But $x = S(T+s)y$ for some $y \in B_\rho^V$ and thus $S(t-s)x = S(t-s + T+s)y = S(T+t)y \in B_\rho^V$ since $S(T+t)B_\rho^V \subset B_\rho^V$. Thus $S(t)x \in \bigcap_{s>0} S(s)B_\rho^V$, that is $S(t)X \subset X$. Reciprocally, if $x \in X$ then

$x = S(t)y$ with some $y \in B_\rho^V$. We want to prove that actually $y \in X$. In order to check this we take $s > 0$ and find $z \in B_\rho^V$ such that $S(t+s)z = x$. This means that $S(t)S(s)z = S(t)y$. From the injectivity property of $S(t)$ it follows that $y = S(s)z$ and thus $y \in S(s)B_\rho^V$. Since s was arbitrary, $y \in X$ and thus $S(t)X \supset X$.

Next we claim that if $Z \subset H$ is bounded and $S(t)Z = Z$ for all $t \geq 0$ it follows that $Z \subset X$. Indeed, if Z is bounded in H there exits t_Z such that $S(t)u_0 \in B_\rho^V$ for all $t \geq t_Z$, $u_0 \in Z$. Then, let $u_0 \in Z$ be arbitrary. We want to show that for every $s > 0$, $u_0 = S(s)y$ for some $y \in B_\rho^V$. From the invariance property of Z it follows that $u_0 = S(s + t_Z)z$ with $z \in Z$. Thus $u_0 = S(s)(S(t_Z)z)$ and $y = S(t_Z)z$ belongs to B_ρ^V. This proves that $u_0 \in \bigcap_{s>0} S(s)B_\rho^V$ and thus that $u_0 \in X$. Let us now take $u_0 \in H$, arbitrary. Let us define $\omega(u_0)$ by

$$\omega(u_0) = \{u \in H \mid \text{there exists } s_j \to \infty \text{ such that}$$

$$u = \lim_{j\to\infty} S(s_j)u_0, \text{ the limit being taken in } H\}.$$

First since $S(t)u_0 \in B_\rho^V$ for $t \geq t_0(|u_0|)$, $\omega(u_0)$ is nonempty and bounded. For each $t \geq 0$, $S(t)(\omega(u_0)) = \omega(u_0)$. Indeed, if $u \in \omega(u_0)$ then $u = \lim_j S(s_j)u_0$ and therefore $S(t)u = \lim_{j\to\infty} S(t+s_j)u_0$. Reciprocally if $u \in \omega(u_0)$ and $S(s_j)u_0$ tend to u then we consider the sequence $S(s_j - t)u_0$ for $s_j \geq t$. Since B_ρ^V is compact and since $S(s_j-t)u_0 \in B_\rho^V$ for all but finitely many j's it follows that, passing to a subsequence $S(s_{j_k} - t)u_0$ converge to an element $v \in B_\rho^V$. Now clearly $S(t)(S(s_{j_k} - t)u_0) = S(s_{j_k})u_0$ converges to u and $S(t)v$ simultaneously, thus $u \in S(t)(\omega(u_0))$.

Proposition 14.1. Let $S(t)$ satisfy properties (i)-(vi) of Proposition 13.1. Let

$$X = \bigcap_{t>0} S(t)B_\rho^V$$

Then

(i) X is compact in H

(ii) $S(t)X = X$ for all $t \geq 0$

(iii) If Z is bounded in H and satisfies $S(t)Z = Z$ for all $t \geq 0$ then $Z \subset X$.

(iv) For every $u_0 \in H$

$$\lim_{t \to \infty} \mathrm{dist}(S(t)u_0, X) = 0$$

(v) X is connected.

Proof. The claims (ii) and (iii) were proven above; (iv) follows from (iii) applied to $Z = \omega(u_0)$. Indeed, if $\mathrm{dist}(S(t_j)u_0, X) \geq \varepsilon > 0$ for a sequence $t_j \to \infty$, then because of the compactness of B_ρ^V the sequence $S(t_j)u_0$ will have a converging subsequence, defining an element of $\omega(u_0)$ which would have to lie outside X, absurd. Property (i) follows form the fact that B_ρ^V and thus $S(t)B_\rho^V$ are all compact. For the proof of the fact that X is connected we reason by contradiction. Assume that D_1 and D_2 are two open (in H) disjoint sets such that $X \subset D_1 \cup D_2$. Assume $x_1 \in X \cap D_1$ and $x_2 \in X \cap D_2$. Let $t > 0$ be arbitrary. Then there exist $y_1 = y_1(t)$, $y_2 = y_2(t)$ in B_ρ^V such that $x_1 = S(t)y_1$, $x_2 = S(t)y_2$. Let us join y_1 to y_2 in B_ρ^V by a straight line, γ. The image under $S(t)$ of γ, $S(t)\gamma$, is a continuous curve joining x_1 to x_2. Therefore there exists at least one point on it which is neither in D_1 nor in D_2. Let us denote it by $x(t) = S(t)y(t)$, $y(t) \in \gamma \subset B_\rho^V$, $x(t) \in H \setminus (D_1 \cup D_2) = F$. The set F is closed and $F \cap X = \phi$. Since for t large $x(t) \in B_\rho^V$ ($S(t)B_\rho^V \subset B_\rho^V$ for t larger than T) there then exist $t_j \to \infty$ such that $x(t_j)$ is convergent in H to some x. Clearly, since $x(t_j) \in F$ and F is closed, $x \in F$. We claim $x \in X$.

Indeed, let $s > 0$ be arbitrary. Take the sequence $S(t_j - s)y(t_j)$, for $t_j \geq s+T$. Since $y(t_j) \in \gamma \subset B_\rho^V$ it follows that $S(t_j - s)y(t_j) \in B_\rho^V$ for $t_j \geq s+T$. Since B_ρ^V is compact there exists a subsequenct $t_{j_k} \to \infty$ such that $S(t_{j_k} - s)y(t_{j_k})$ converges to some $y \in B_\rho^V$. Thus $S(s)(S(t_{j_k} - s)y(t_{j_k})$ converges to $S(s)y$ and $S(s)(S(t_{j_k}-s)y(t_{j_k})) = S(t_{j_k})y(t_{j_k})$ converges to x as a subsequence of $S(t_j)y(t_j)$. Thus $x = S(s)y$ with $y \in B_\rho^V$, for arbitrary $s > 0$ and therefore $x \in X$. This is absurd since $x \in X \cap F$ and $X \cap F$ is empty.

Definition 14.2. The set X is called the universal attractor of the equation (13.1).

We introduce now global Lyapunov exponents ([CF1]). Let $t > 0$, $u_0 \in V$. We define the linear operator $S'(t,u_0):H \to H$ by

(14.2) $$S'(t,u_0)\xi = v(t,u_0,\xi)$$

where $v(t,u_0,\xi)$ is the solution of

(14.3) $$\frac{dv}{dt} + \nu Av + B(S(t)u_0,v) + B(v,S(t)u_0) = 0$$

(14.4) $$v(0) = \xi$$

computed at t.

From Proposition 13.2 it follows that $S'(t,u_0):H \to V$ is bounded and therefore $S'(t,u_0)$ is a compact operator in H. Let us denote by

(14.5) $$M(t,u_0) = [(S'(t,u_0))^* S'(t,u_0)]^{1/2}$$

$M(t,u_0)$ is compact selfadjoint, nonnegative. We denote by $m_j(t,u_0)$ the eigenvalues of M counted according to their multiplicities:

$$0 < \ldots \leq m_N(t,u_0) \leq m_{N-1}(t,u_0) \leq \ldots \leq m_1(t,u_0).$$

The numbers $m_j(t,u_0)$ are called the singular values of $S'(t,u_0)$. Let us consider an orthonormal family $\phi_j(t,u_0)$ of eigenvectors for $M(t,u_0)$ corresponding to the eigenvalues $m_j(t,u_0)$. From the uniqueness of solutions of (14.3) and the semigroup property of $S(t)$ it follows that

$$(14.6)\qquad S'(t+s,u_0)\xi = S'(t,S(s)u_0)S'(s,u_0)\xi$$

for all $\xi \in H$, $t,s \geq 0$. Therefore, if $S'(s,u_0)\xi = 0$ for some $s > 0$ it will remain zero for all $\tau \geq s$. From property (i) of Proposition 14.2, it follows that $S'(t,u_0)\xi = 0$ for all $t > 0$ and from strong continuity $\xi = 0$. Therefore, $M(t,u_0)$ is injective for every $t > 0$ and $\phi_j(t,u_0)$ form a basis. Let us denote by $\psi_j(t,u_0)$ the vectors

$$(14.7)\qquad \psi_j(t,u_0) = S'(t,u_0)\phi_j(t,u_0).$$

Then

$$(14.8)\qquad (\psi_j,\psi_k) = \delta_{jk}m_jm_k\ .$$

(We will omit the dependence of ϕ_j, ψ_j, m_j on t and u_0 when no confusion can arise from the omission.) We obtain thus a representation of $S'(t,u_0)$ as the sum

$$(14.9)\qquad S'(t,u_0)\xi = \sum_{j=1}^{\infty}(\phi_j,\xi)\psi_j.$$

Lemma 14.3. There exists a positive continuous function $c(t)$ defined for every $t > 0$, depending on ν, $|f|$ and ρ such that, for every $u_1,u_0 \in B_\rho^V$ the following estimate holds

$$(14.10)\quad |S(t)u_1 - S(t)u_0 - S'(t,u_0)(u_1 - u_0)| \leq c(t)|u_1 - u_0|^2.$$

Proof. One considers the difference $w(t) = S(t)u_1 - S(t)u_0 - S'(t,u_0)(u_1 - u_0)$ and uses the first energy equation for $w(t)$. We omit further details.

The inequality (14.10) implies that $S'(t,u_0)$ is the Frechet derivative $\frac{DS(t)}{D(u_0)}$ of $S(t)$.

Lemma 14.4. Let $u_1, u_0 \in B_\rho^V$ and let $N \geq 1$ be an integer. Then

$$(14.11) \quad |S(t)u_1 - [S(t)u_0 + \sum_{j=1}^{N} (\phi_j(t,u_0), u_1 - u_0)\psi_j(t,u_0)]| \leq (m_{N+1}(t,u_0) + c(t)|u_1 - u_0|)|u_1 - u_0|$$

Proof. Combine (14.10) with $|\sum_{j=N+1}^{\infty} (\phi_j, \xi)\psi_j|^2 \leq m_{N+1}^2 |\xi|^2$.

The geometrical interpretation of (14.9) and (14.10) is that, up to an error of order r^2, $S(t)$ transforms a ball in H centered at u_0 and of radius r into an infinite dimensional ellipsoid, centered at $S(t)u_0$ and with semi axes on the directions $\psi_j(t,u_0)$ and of lengths $rm_j(t,u_0)$. We consider the N dimensional ellipsoid

$$(14.12) \quad \mathcal{E}_N(t,u_0,r) = \{v | v = S(t)u_0 + \sum_{j=1}^{N} (\phi_j(t,u_0),\xi)\psi_j(t,u_0),\ |\xi| < r\}.$$

Its N dimensional volume is less than the volume of the corresponding box:

$$(14.13) \quad \mathrm{vol}_N(\mathcal{E}_N(t,u_0,r)) \leq (2r)^N m_1(t,u_0) \ldots m_N(t,u_0).$$

The classical Lyapunov exponents are numbers $\mu_j(u_0)$ such that, asymptotically $m_j(t,u_0) \sim \exp t\mu_j(u_0)$. We want to define global Lyapunov exponents.

Let $P_N(t,u_0)$ denote the number

(14.14) $P_N(t,u_0) = m_1(t,u_0) \dots m_N(t,u_0).$

We observe that

(14.15) $$P_N(t,u_0) = \sup_{\substack{\xi_i \in H, i=1,..,N \\ |\xi_i| \le 1}} |S'(t,u_0)\xi_1 \wedge \dots \wedge S'(t,u_0)\xi_N|$$

Indeed

$$\begin{aligned} |S'(t,u_0)\xi_1 \wedge \dots \wedge S'(t,u_0)\xi_N|^2 &= \det_{i,j} (S'(t,u_0)\xi_i, S'(t,u_0)\xi_j) \\ &= \det_{i,j} (M(t,u_0)\xi_i, M(t,u_0)\xi_j) \\ &= |M(t,u_0)\xi_1 \wedge \dots \wedge M(t,u_0)\xi_N|^2 \\ &\le m_1^2(t,u_0) \wedge \dots \wedge m_N^2(t,u_0) |\xi_1 \wedge \dots \wedge \xi_N|^2. \end{aligned}$$

The supremum is achieved for $\xi_i = \phi_i(t,u_0)$. It follows from (14.15) and (14.6) that

(14.16) $P_N(t+s,u_0) \le P_N(t,S(s)u_0)P_N(s,u_0).$

Let us set

(14.17) $$P_N(t) = \sup_{u_0 \in B_\rho^V} P_N(t,u_0)$$

(14.18) $$\overline{m}_j(t) = \sup_{u_0 \in B_\rho^V} m_j(t,u_0).$$

the numbers $P_N(t)$ and $\overline{m}_j(t)$ are finite

$$P_N(t) \le \overline{m}_1(t) \dots \overline{m}_N(t) \le \overline{m}_1(t)^N$$

$$\overline{m}_j(t) \le \overline{m}_1(t).$$

The fact that $\overline{m}_1(t)$ is finite is a consequence of Proposition 13.2, (ii); $\overline{m}_1(t) \le \exp(k_1 \nu \lambda_1 t)$.

From (14.16) it follows that the function $\log P_N(t)$ is subadditive. The numbers $P_N(t)$, $\overline{m}_j(t)$ are never zero because $M(t,u_0)$ are injective.

Definition 14.5. Let $j \geq 1$ be an integer. We define global Lyapunov exponents μ_j, $\overline{\mu}_j$ by

(14.19) $$\overline{\mu}_j = \limsup_{t \to \infty} \frac{1}{t} \log \overline{m}_j(t)$$

(14.20) $$\pi_j = \lim_{t \to \infty} \frac{1}{t} \log P_j(t)$$

(14.21) $$\begin{cases} \mu_1 = \pi_1, & \text{and inductively} \\ \mu_{j+1} = \pi_{j+1} - \pi_j, & j \geq 1 \end{cases}$$

Note that, from these definitions it follows that

(14.22) $$\mu_j \leq \overline{\mu}_j \leq \frac{\mu_1 + \dots + \mu_j}{j} \qquad j \geq 1$$

and clearly $\mu_1 + \dots + \mu_j = \pi_j$.

Proposition 14.6. For every N there exists a positive continuous function $c_N(t)$ defined for $t > 0$ such that, for $u_1, u_2 \in B_\rho^V$

(14.23) $$|P_N^2(t,u_1) - P_N^2(t,u_2)| \leq c_N(t)|u_1 - u_2|^{1/2}.$$

Proof. Since the linear operator $M(t,u_j)$ $j = 1,2$ have norms (as operators in H) bounded by $e^{k_1\nu\lambda_1 t}$ (Proposition 13.2,(ii)) it follows that, for $|\xi_i| \leq 1$, $i = 1,\dots,N$,

$$|\det_{i,j}(M(t,u_1)\xi_i,M(t,u_1)\xi_j) - \det_{i,j}(M(t,u_2)\xi_i,M(t,u_2)\xi_j)|$$
$$\leq \gamma_N(t) \; \| M(t,u_1) - M(t,u_2) \|_{\mathcal{L}(H,H)}$$

where $\gamma_N(t)$ is of the form $\gamma_N e^{N\nu\lambda_1 k_1 t}$ with γ_N a constant.

Using the representation (14.15) we deduce that

$$|P_N^2(t,u_1) - P_N^2(t,u_2)| \leq \gamma_N(t) \; \|M(t,u_1) - M(t,u_2)\|_{\mathcal{L}(H,H)}.$$

Now for any two bounded, nonnegative, selfadjoint operators M_1 and M_2

$$\| M_1 - M_2 \| \leq \frac{4\sqrt{2}}{\pi} \| M_1^2 - M_2^2 \|^{1/2}$$

where $\| \; \|$ means the norm as an operator. It follows that

$$\| M(t,u_1) - M(t,u_2) \|_{\mathcal{L}(H,H)} \leq \frac{8}{\pi} e^{k_1\nu\lambda_1 t/2} \|S'(t,u_1) - S'(t,u_2)\|^{1/2}_{\mathcal{L}(H,H)}$$

Finally, we estimate

$$\|S'(t,u_1) - S'(t,u_2)\|_{\mathcal{L}(H,H)} \leq C(t)|u_1 - u_2|$$

by standard energy estimates.

As a consequence we have

Lemma 14.7. For every $t > 0$, $N \geq 1$, let

$$\underline{m}_N(t) = \inf_{u_0 \; B_\rho^V} m_N(t,u_0)$$

The number $\underline{m}_N(t)$ is strictly positive

$$\underline{m}_N(t) > 0$$

Proof. Indeed, assume that for some $N \geq 1$ and $t > 0$, $\underline{m}_N(t) = 0$. Then, in view of $P_N^2(t,u_0) \leq e^{2(N-1)k_1\nu\lambda_1 t} m_N^2(t,u_0)$, the compactness of B_ρ^V and the continuity of $P_N(t,.)$ there would exist $u_0 \in B_\rho^V$ such that $P_N^2(t,u_0) = 0$.

But this implies $m_j(t,u_0) = 0$ for some j, $1 \le j \le N$, contradicting the fact that $M(t,u_0)$ is injective.

In the language of global Lyapunov exponents μ_j, $\bar{\mu}_j$ the statements of exponential decay of volume elements of Section 13 become negative upper bounds for $\mu_1 + \cdots + \mu_j$ for j large. For instance, from Theorem 13.6 we infer that

$$(14.24) \qquad \bar{\mu}_j \le \frac{\mu_1 + \cdots + \mu_j}{j} \le -\frac{c_0}{4}\nu\lambda_1$$

for $j \ge c_4 G$, j positive integer. (See (13.40).)

Actually, in the proof of Theorem 13.6 we showed that

$$|v_1(t) \wedge \cdots \wedge v_N(t)|^2 \le |v_1(0) \wedge \cdots \wedge v_N(0)|^2 \exp(-t\tilde{\gamma}_N)$$

for $t \ge t_0(\rho)$ with $\tilde{\gamma}_N = (\frac{c_0}{4} N(N+1) - c_L G^2)\nu\lambda_1$. In view of (14.15), the definition of $P_N(t)$, π_N and μ_i we deduce that in the two dimensional Navier-Stokes case

$$(14.25) \qquad j\bar{\mu}_j \le \mu_1 + \cdots + \mu_j \le \nu\lambda_1(c_L G^2 - \frac{c_0}{4} j(j+1))$$

for all $j = 1,2,\ldots$.

Estimates of the type of (14.25) are valid for the periodic two dimensional case too. Defining the Lyapunov exponents in V instead of H would improve, in the periodic case, estimates (14.24) and (14.25).

Let us recal the definitions ((10.20), (10.21)) of Hausdorff dimension and define the fractal dimension. Let $Z \subset H$ be a compact set. The Hausdorff dimension of Z is

$$(14.26) \qquad d_H(Z) = \inf\{D > 0 \mid \mu_H^D(Z) = 0\}$$

where

$$(14.27) \qquad \mu_H^D(Z) = \lim_{r \downarrow 0} \mu_{H,r}^D(Z)$$

and

$$(14.28) \quad \mu^D_{H,r}(Z) = \inf\{\sum_{i=1}^{k} r_i^D \mid Z \subset \bigcup_{i=1}^{k} B_i,\ B_i \text{ balls in } H \text{ of radii } r_i \le r\}.$$

The fractal dimension of Z is defined as

$$(14.29) \quad d_M(Z) = \limsup_{r \to 0} \frac{\log n_Z(r)}{\log 1/r}$$

where

$$(14.30) \quad n_Z(r) = \text{minimal number of balls in } H \text{ of radii} \le r \text{ needed to cover } Z.$$

Let us observe that an alternative and equivalent definition of $d_M(Z)$ is

$$(14.31) \quad d_M(Z) = \inf\{D > 0 \mid \mu^D_M(Z) = 0\}$$

where

$$(14.32) \quad \mu^D_M(Z) = \limsup_{r \to 0} r^D n_Z(r).$$

This second way of defining $d_M(Z)$ shows clearly the difference between Hausdorff and fractal dimension. First

$$(14.33) \quad d_H(Z) \le d_M(Z)$$

because, for each $r > 0$

$$(14.34) \quad \mu^D_{H,r}(Z) \le r^D n_Z(r)$$

However, the inequality (14.33) can be strict. Actually, there are examples of compact sets $Z \subset H$ with $d_H(Z) = 0$ and $d_M(Z) = \infty$! ([CF1]). The difference originates from the fact that for fractal dimension a fine cover $Z \subset \bigcup B_i$ with B_i balls of radii $r_i \le r$, has the same weight in the computation as the coarser cover when all the balls B_i are dilated to have radius r.

The method of estimate of both the Hausdorff and the fractal dimensions that we will use is roughly the following. Suppose the set X is covered by a finite number of balls of radius less than r. Let S(t) transport each ball for a very long time. Then each ball becomes a slightly distorted ellipsoid whose semiaxes have lengths of of the size $m_j(t,u_0)r$. Cover these ellipsoids by smaller balls. In this process of covering again, the control on the volumes implies control on the number of balls.

Lemma 14.8. Let $\mathcal{E}_N$ be an N dimensional ellipsoid with semiaxes of lengths rm_j, $j = 1,\dots,N$ where $r > 0$ and $m_1 \geq m_2 \geq \dots \geq m_N > 0$. Let a be a positive number. Then the number of balls of radius ra needed to cover $\mathcal{E}_N$ does not exceed

$$(14.35) \quad (2\sqrt{N})^{\ell} \cdot \frac{m_1 \cdots m_\ell}{a^\ell}$$

where ℓ is the largest number $1 \leq \ell \leq N$ such that $\frac{a}{\sqrt{N}} < m_\ell$. If $\frac{a}{\sqrt{N}} \geq m_1$ then one ball suffices.

Proof. The ellipsoid $\mathcal{E}_N$ is included in a box whose sides have lengths $2rm_j$, $j = 1,\dots,N$. We will cover this box with N dimensional cubes of side lengths $\frac{2ra}{\sqrt{N}}$. Then each such cube will be contained in a ball of radius ra. We count how many cubes we need.

Each side of the box will be divided in $[\frac{2rm_j}{\frac{2ra}{\sqrt{N}}}] + 2$ subdivision points where [ ] denotes integer part, yielding

$$([\sqrt{N}\frac{m_1}{a}] + 1)([\sqrt{N}\frac{m_2}{a}] + 1) \dots ([\sqrt{N}\frac{m_N}{a}] + 1)$$

cubes. In this product only the factors where $\sqrt{N}\frac{m_j}{a} \geq 1$ contribute. For such j we majorize $[\sqrt{N}\frac{m_j}{a}] + 1$ by $2\sqrt{N}\frac{m_j}{a}$. This yields (14.35).

Let $u_0 \in X$ where X is the universal attractor (or any other bounded invariant set, for that matter; however since the universal attractor is the largest such set, upper bounds on its dimensions are upper bounds for the dimensions of any bounded invariant set). Let $B_H(u_0,r)$ denote a ball in H of radius r around u_0. Let $t > 0$ be a large time, to be fixed later. Let us consider $S(t)B_H(u_0,r)$. Let $\mathcal{E}_N(t,u_0,r)$ be the N dimensional ellipsoid defined in (14.12). The estimate (14.11) of Lemma 14.4 implies that the distance between any point $S(t)u_1$ of $S(t)B_H(u_0,r)$ to $\mathcal{E}_N(t,u_0,r)$ is bounded by

(14.36) $$\text{dist}(S(t)u_1, \mathcal{E}_N(t,u_0,r)) \leq (m_{N+1}(t,u_0) + c(t)r)r.$$

Let us call θ the stretching factor in (14.36)

(14.37) $$\theta = \theta(t,u_0,r) = m_{N+1}(t,u_0) + c(t)r.$$

Lemma 14.9. Let N be a positive integer and D a positive number satisfying

(i) $\bar{\mu}_{N+1} < 0$

(ii) $N \leq D \leq N+1$

(iii) $(D - N)\mu_{N+1} + \mu_1 + \dots + \mu_N < 0$

Then $\mu_H^D(X) = 0$.

Theorem 14.10 (Kaplan-Yorke formula). Let

(14.38) $$j_0 = \max\{j \mid \mu_1 + \dots + \mu_j \geq 0\}$$

Then

(14.39) $$d_H(X) \leq j_0 + \frac{\mu_1 + \dots + \mu_{j_0}}{|\mu_{j_0+1}|}$$

Proof. If $D \geq j_0+1$ then we can apply Lemma 14.9 with $N = [D]$. Assumption (iii) is true because $\mu_1 + \dots + \mu_N < 0$, $\mu_{N+1} < 0$. If D satisfies

$$j_0 + \frac{\mu_1 + \dots + \mu_{j_0}}{|\mu_{j_0+1}|} < D \leq j_0 + 1$$

we can apply Lemma 14.9 with $N = j_0$. Since $\bar{\mu}_{N+1} \leq \frac{\mu_1 + \dots + \mu_{N+1}}{N+1}$ and by the definition of j_0, $\mu_1 + \dots + \mu_{j_0+1} < 0$ we check (i). Assumption (iii) is valid because $|\mu_{j_0+1}| = -\mu_{j_0+1}$. Thus $\mu_H^D(X) = 0$ for $j_0 + \frac{\mu_1 + \dots + \mu_{j_0}}{|\mu_{j_0+1}|} < D$ and and therefore (14.39) is true.

Theorem 14.11.

(a) In the two dimensional Dirichlet boundary conditions case

(14.40) $d_H(X) \leq c_4 G + 1$

(b) In the two dimensional periodic boundary conditions case

(14.41) $d_H(X) \leq c_6(1 + G)^{2/3}(\log (G+2))^{1/3}$

Proof. The inequality $\mu_1 + \dots + \mu_{j_0} + \mu_{j_0+1} < 0$ implies $\frac{\mu_1 + \dots + \mu_{j_0}}{|\mu_{j_0+1}|} \leq 1$ since μ_{j_0+1} is negative. Thus, from (14.39),

(14.42) $d_H(X) \leq j_0 + 1$.

Now (14.40) and (14.41) are obtained from (14.42) and the upper bounds (13.39) and (13.67) for j_0+1. In the periodic case all operations are in V. A simple lemma is required, to prove that the dimension of X when computed in V is the same as when computed in H.

Proof of Lemma 14.9. Let us take $\varepsilon > 0$ such that

$$(14.43) \quad (D - N)\mu_{N+1} + \mu_1 + \cdots + \mu_N + 3\varepsilon < 0.$$

Let us choose t large enough to satisfy

$$(14.44) \quad \frac{1}{2} N \log N + (N + 1)\log 2 + D \log 3 \leq \varepsilon t$$

$$(14.45) \quad e^{-\varepsilon t} \leq 1/2$$

$$(14.46) \quad \overline{m}_{N+1}(t) \leq 1/8$$

$$(14.47) \quad \begin{cases} \dfrac{\log P_N(t)}{t} \leq \mu_1 + \cdots + \mu_N + \varepsilon \\ \dfrac{\log P_{N+1}(t)}{t} \leq \mu_1 + \cdots + \mu_{N+1} + \varepsilon \end{cases}$$

The requirements (14.44), (14.45) are only largeness assumptions. The requirement (14.47) can be achieved in view of the definition of the global Lyapunov exponents μ_i. Requirement (14.46) follows from assumption (i) of the lemma. We fix t with these properties. We choose now $r_0 > 0$ such that

$$(14.48) \quad 1 + \frac{c(t)r_0}{\underline{m}_{N+1}(t)} \leq 2.$$

Recall that $\underline{m}_{N+1}(t) = \inf_{u_0 \in B_\rho^V} m_{N+1}(t,u_0) > 0$. (Lemma 14.7). From the definition (14.37) for the stretching factor $\theta = \theta(t,u_0,r)$ it follows that for any $u_0 \in B_\rho^V$ and $r \leq r_0$

$$(14.49) \quad \theta(t,u_0,r) \leq 2m_{N+1}(t,u_0).$$

Let us consider the image $S(t)B_H(u_0,r)$ of a ball of radius $r \leq r_0$ centered at $u_0 \in X$. Since the distance from $S(t)B_H(u_0,r)$ to $\mathcal{E}_N(t,u_0,r)$ is not larger than $2m_{N+1}(t,u_0)r$ ((14.36), (14.37), (14.49)) if follows that a covering of $\mathcal{E}_N(t,u_0,r)$ with balls of radii $rm_{N+1}(t,u_0)$ will yield by dilation a covering with balls of same centers and radii $3rm_{N+1}(t,u_0)$ of $S(t)B_H(u_0,r)$. Now by Lemma 14.8 with $a = m_{N+1}(t,u_0)$ we obtain that

(14.50) the number of balls of radii $3m_{N+1}(t,u_0)r$ needed to cover $S(t)B_H(u_0,r)$ does not exceed

$$(2\sqrt{N})^N \frac{P_N(t,u_0)}{(m_{N+1}(t,u_0))^N}.$$

Note that the radii $3m_{N+1}(t,u_0)r$ are not larger than $r/2$ by (14.46). Now let us take a finite cover of X by balls $B_H(u_i,r_i)$, $1 \le i \le k$, $r_i \le r \le r_0$. From the invariance property of X

$$X \subset \bigcup_{i=1}^{k} S(t)B_H(u_i,r_i).$$

We cover each set $S(t)B_H(u_i,r_i)$ with balls of radii $3m_{N+1}(t,u_i)r_i \le \frac{r_i}{2} \le \frac{r}{2}$. We obtain a new cover of X with balls of radii not larger than $\frac{r}{2}$. Then we compute $\mu^D_{H,\frac{r}{2}}(X)$:

$$\mu^D_{H,\frac{r}{2}}(X) \le \sum_{i=1}^{k} (2\sqrt{N})^N \frac{P_N(t,u_i)}{(m_{N+1}(t,u_i))^N} (3r_i)^D (m_{N+1}(t,u_i))^D. \tag{14.51}$$

Now we write $P_N(t,u_i)(m_{N+1}(t,u_i))^{D-N}$ as

$$P_N(t,u_i)(m_{N+1}(t,u_i))^{D-N} = P_N(t,u_i)^{1+N-D}P_{N+1}(t,u_i)^{D-N}.$$

The exponents 1+N-D and D-N are both nonnegative and thus from (14.47) it follows that

$$P_N(t,u_i)^{1+N-D}P_{N+1}(t,u_i)^{D-N} \le \exp((D-N)\mu_{N+1} + \mu_1 + \dots + \mu_N + \varepsilon)t) \le \exp(-2\varepsilon t)$$

in view of (14.43).

On the other hand, the numerical constant in (14.51), $(2\sqrt{N})^{N_3 D}$ is majorized by

$$(2\sqrt{N})^{N_3 D} \leq e^{\varepsilon t}$$

because of (14.44).

Combining we get, in view of (14.45)

$$\mu^D_{H,\frac{r}{2}}(X) \leq e^{-\varepsilon t} \sum_{i=1}^{k} r_i^D \leq \frac{1}{2} \sum_{i=1}^{k} r_i^D .$$

Since the cover $B_H(u_i, r_i)$ was arbitrary we obtained

$$\mu^D_{H,\frac{r}{2}}(X) \leq \frac{1}{2} \mu^D_{H,r}(X)$$

Since the function $r \to \mu^D_{H,r}(X)$ is nonincreasing it follows that $\mu^D_H(X) = 0$. The proof of Lemma 14.9 is complete.

We pass now to the fractal dimension.

Lemma 14.12. Let N be an integer and D a real number. Assume

(i) $D \geq N$

(ii) $(D - \ell)\bar{\mu}_{N+1} + \mu_1 + \dots + \mu_\ell < 0$ for all $\ell = 1,2,\dots,N$

(iii) $\bar{\mu}_{N+1} < 0$

Then $\mu^D_M(X) = 0$.

Proof. Let us choose $\varepsilon > 0$ such that

(14.52) $\quad (D - \ell)\bar{\mu}_{N+1} + \mu_1 + \dots + \mu_\ell + (D - \ell + 3)\varepsilon \leq 0$

for all $\ell = 0, 1, 2,\dots,N$. (For $\ell = 0$, $\mu_1 + \dots + \mu_\ell$ is taken = 0).

We choose t large enough to insure the validity of (14.44)-(14.46) and also

$$(14.53)\quad \begin{cases} \dfrac{\log \overline{m}_{N+1}(t)}{t} \le \overline{\mu}_{N+1} + \varepsilon \\ \dfrac{\log P_\ell(t)}{t} \le \mu_1 + \cdots + \mu_\ell + \varepsilon \text{ for } \ell = 1,\ldots,N. \end{cases}$$

Let $u_0 \in X$ and $B_H(u_0,r)$ with $r \le r_0$, r_0 defined in(14.48). We take $S(t)B_H(u_0,r)$ and cover it with balls of radius $3\overline{m}_{N+1}(t)r$. Applying Lemma 14.8, with $a = \overline{m}_{N+1}(t)$ we deduce that

(14.54) The number of balls of radii $3\overline{m}_{N+1}(t)r$ needed to cover $S(t)B_H(u_0,r)$ does not exceed

$$\max\{(2\sqrt{N})^N(\overline{m}_{N+1}(t))^{-\ell}P_\ell(t),1\}$$

for some $\ell = \ell(N,u_0)$, $1 \le \ell \le N$. Considering the minimal number of balls of radii $\le 3\overline{m}_{N+1}(t)r$ needed to cover X we obtain the relation

$$n_X(3\overline{m}_{N+1}(t)r) \le [(2\sqrt{N})^N \max_{1<\ell<N} ((\overline{m}_{N+1}(t)^{-\ell}P_\ell(t))]n_X(r)$$

Multiplying by $(3\overline{m}_{N+1}(t)r)^D$ we get

$$(3\overline{m}_{N+1}(t)r)^D n_X(3\overline{m}_{N+1}(t)r) \le [3^D(2\sqrt{N})^N \max_{1<\ell<N} (\overline{m}_{N+1}(t))^{D-\ell}P_\ell(t)]n_X(r)r^D$$

Now the quantity in square brackets is less than 1/2.

The function $\varphi(r) = r^D n_X(r)$ satisfies

(a) $\sup\{\varphi(r) \mid 3\overline{m}_{N+1}(t)r_0 \le r \le r_0\} < \infty$

(b) $\varphi(3\overline{m}_{N+1}(t)r) \le \frac{1}{2}\varphi(r)$, for all $r \le r_0$.

It is elementary to check that such a function $\varphi(r)$ must have

$$\lim_{r \to 0} \varphi(r) = 0.$$

This proves the lemma.

Theorem 14.13. Let j_0 be defined as in (14.38)

$$j_0 = \mathrm{Max}\{j \mid \mu_1 + \dots + \mu_j \geq 0\}.$$

Then

$$\text{(14.55)} \quad d_M(X) \leq \max_{1 \leq \ell \leq j_0} \left(\ell + \frac{\mu_1 + \dots + \mu_\ell}{|\bar{\mu}_{j_0+1}|}\right)$$

Proof. If $D > \max_{1 \leq \ell \leq j_0} \left(\ell + \frac{\mu_1 + \dots + \mu_\ell}{|\bar{\mu}_{j_0+1}|}\right)$ Then $D > j_0$ because

$$\mu_1 + \dots + \mu_{j_0} \geq 0. \quad (D - \ell)\bar{\mu}_{j_0+1} + \mu_1 + \dots + \mu_\ell < 0$$

for all $\ell = 1,\dots,j_0$, because $\bar{\mu}_{j_0+1}$ is negative. Therefore the conclusion of the theorem follows directly from Lemma 14.12, applied with $N = j_0$.

Now, from (14.25) we know that

$$\ell\bar{\mu}_\ell \leq \mu_1 + \dots + \mu_\ell \leq \nu\lambda_1(c_L G^2 - \frac{c_0}{4}\ell(\ell+1)),$$

for all ℓ.

Assume N is such that

$$\text{(14.56)} \quad \frac{c_0}{8}(N+2)(N+1) > c_L G^2.$$

It follows that

$$\text{(14.57)} \quad \bar{\mu}_{N+1} \leq -\nu\lambda_1(\frac{c_0}{8}(N+2)).$$

Take $D \geq 2N+2$. Then

$$(D - \ell)\bar{\mu}_{N+1} + (\mu_1 + \dots + \mu_\ell) \leq -\nu\lambda_1 \frac{c_0}{8}(N+1)(N+2) + \nu\lambda_1 c_L G^2 < 0$$

for all $\ell = 1,2,\dots,N+1$.

We can apply Lemma 14.12 for N+1:

Theorem 14.14.

(a) In the two dimensional Dirichlet boundary conditions case

$$d_M(X) \leq c_8 G + 1$$

(b) In the two dimensional periodic boundary conditions case

$$d_M(X) \leq c_9 (G + 1)^{2/3} (\log(G + 2))^{1/3}.$$

For the three dimensional Navier-Stokes equations the previous results work under the assumption of regularity. For instance, if Z is a bounded invariant set in V then one can define the quantity (see (13.72))

$$(14.58) \quad \eta = \limsup_{t \to \infty} \left(\sup_{u_0 \in Z} \frac{1}{\nu^4 \lambda_1} \frac{1}{t} \int_0^t \| S(s) u_0 \|^4 ds \right)$$

Applying Theorem 13.8 and the Lemmas 14.9, 14.12 one can prove (see [CFT1])

Theorem 14.15. Let $Z \subset V$ be bounded. Assume Z invariant. Then

$$(14.59) \quad d_H(Z) \leq d_M(Z) \leq c_9 (1 + \eta)^{3/2}.$$

The above bounds can be understood in the context of the traditional estimates of the number of degrees of freedom of turbulent flows ([L-L], [K]). This number is

$$(14.60) \quad N \sim \left(\frac{\ell_0}{\ell_c}\right)^d$$

where ℓ_0 is the linear size of the region occupied by the fluid, d = 2 or 3 is the spatial dimension and ℓ_c is a small scale, below which viscosity effects determine entirely the motion. Thus $N = \left(\frac{\ell_0}{\ell_c}\right)^d$ is simply an account of the number of mesh points dividing a cube of side length ℓ_0 into divisions of length ℓ_c. The scale ℓ_c is defined differently for d = 2 and d = 3. If d = 3 then ℓ_c is defined via an energy dissipation flux

$$(14.61)\qquad \varepsilon = \nu \langle \nabla u \rangle^2$$

where < > denotes ensemble averaging. By dimensional analysis the only length one can form with ν, ε is

$$(14.62)\qquad \ell_c = \frac{\nu^{3/4}}{\varepsilon^{1/4}} .$$

In the two-dimensional case, d = 2, the role of ε is played by an enstrophy flux

$$(14.63)\qquad \chi = \nu \langle \Delta u \rangle^2 .$$

The only length one can form with ν, χ is

$$(14.64)\qquad \ell_c = \left(\frac{\nu^3}{\chi}\right)^{1/6} .$$

It is very significant that the estimates of the fractal dimension of the attractor in d = 3 (14.59) and d = 2 (13.64) can be expressed in the form (14.60) provided the average operation is defined as a time average on the attractor. For instance, if the quantity

$$(14.65)\qquad \langle \Delta u \rangle = \lambda_1^{1/2} \sup_{u_0 \in X} \limsup_{t \to \infty} \left(\frac{1}{t}\int_0^t |AS(s)u_0|^2 ds\right)^{1/2}$$

(see (13.63)) is taken to represent the average of the Laplacian then (13.64) implies that the fractal dimension of the universal attractor for 2D Navier Stokes equations is bounded by

$$(14.66)\qquad d_M(x) \leq c\left(1 + \left(\frac{\ell_0}{\ell_c}\right)^2\right)\left(1 + \log\left(\frac{\ell_0}{\ell_c}\right) + 1\right)^{1/3}$$

with $\ell_0 = \frac{1}{\sqrt{\lambda_1}}$ and ℓ_c given by (14.64) via (14.63) and (14.65).

For more details see [CFT1] for d = 3 and [CFT2] for d = 2.

15

INERTIAL MANIFOLDS

In this section we study the question: can one imbed the universal attractor X into a finite dimensional, regular manifold Y which is invariant (forward in time) and which attracts exponentially all trajectories? If one could find such Y and if Y would be smooth enough then the restriction of the PDE to Y would be an ODE, called an inertial form of the PDE, whose dynamics would contain all the information about the long time behavior of the solutions to the PDE. The manifold Y is called an inertial manifold. The existence of such inertial manifolds for the Navier-Stokes equations, even in the two dimensional case, is unknown. A certain number of dissipative PDE's do possess inertial manifolds ([F-S-T], [CFNT]). Among them, a number are one dimensional as the Kuramoto-Sivashinski, nonlocal Burgers, Cahn-Hilliard equations are. Reaction diffusion equations can be treated in one and two spatial dimensions. There are, to present, two techniques of constructing inertial manifolds. The one that we present here ([CFNT]) constructs the inertial manifold as an integral manifold starting from an explicit, simple, smooth finite dimensional manifold of initial data and integrating forward in time. The strong dissipative linear principal part of the equation helps to control the time evolution of the position of the tangent space to the integral surface. We will present here the method of [CFNT] illustrated on a simple example,

(15.1) $$\frac{\partial u}{\partial t} - \Delta u + \varepsilon\Delta^2 u + (u\cdot\nabla)u + \nabla p = f$$

(15.2) $$\operatorname{div} u = 0$$

with periodic boundary conditions in $\mathbb{R}^2$.

We chose the example (15.1), (15.2) rather than one of the more physically significant equations for which the same construction works because the notation and background of the Navier-Stokes equations established in the previous sections can be used. The equation (15.1), (15.2) has thus the form

(15.3) $$\frac{du}{dt} + \varepsilon A^2 u + Au + B(u,u) = f$$

(15.4) $$u(0) = 0.$$

We take the two dimensional periodic case; we take the period to be 2π so that $\lambda_1 = 1$. We will give first some bounds for $u(t) = S(t)u_0$. Clearly, since

(15.5) $$(B(u,u),u) = 0$$ and

(15.6) $$(B(u,u),Au) = 0$$

we have

(15.7) $$|u(t)|^2 \leq |u_0|^2 e^{-t} + |f|^2(1 - e^{-t})$$

(15.8) $$\|u(t)\|^2 \leq \|u_0\|^2 e^{-t} + |f|^2(1 - e^{-t}).$$

Then taking the scalar product of (15.3) with A^2u we get

$$\frac{1}{2}\frac{d}{dt}|Au|^2 + \varepsilon|A^2u|^2 \leq \frac{|f|^2}{\varepsilon} + \varepsilon\frac{|A^2u|^2}{4} + |(B(u,u),A^2u)|$$

Using (6.20) with $s_3 = 0$, $s_1 = 1/2$, $s_2 = 1/2$ we get

$$|B(u,u,A^2u)| \le c|u|^{1/2}\|u\|^{1/2}\|u\|^{1/2}|Au|^{1/2}|A^2u|$$
$$\le c|u|^{1/2}\|u\||A^2u|^{3/2} \le \varepsilon \frac{|A^2u|^2}{4} + c_\varepsilon|u|\|u\|^4$$

We obtain

$$(15.9) \qquad |Au(t)|^2 \le |Au_0|^2 + \frac{|f|^2}{\varepsilon^2} + c_\varepsilon(\|u_0\|^2 + |f|^2)^3.$$

We could have gotten a much better dependence of $|f|$ on the right hand side of (15.9) but we do not attempt to optimize results here.

From (15.9) and the estimate

$$(15.10) \qquad |u|_{L^\infty} \le c\|u\|(1 + \log_+|Au|)$$

where $\log_+ \lambda = \log \lambda$ for $\lambda \ge 1$ and $\log_+ \lambda = 0$ for $0 < \lambda < 1$ it follows that

$$(15.11) \quad |u(t)|_{L^\infty}$$
$$\le c(\|u_0\| + |f|)(1 \log_+[|Au_0|^2 + \frac{|f|^2}{\varepsilon^2} + c_\varepsilon(\|u_0\|^2 + |f|^2)^3])$$

First we define and study some objects of independent interest.

Let $Q_0:H \to H$ be a projector of N dimensional range. We think of this range as being the tangent space at some $u_0 \in H$ to an N dimensional surface Σ. We let the solution map $S(t)$ transport this surface. At each $t > 0$ the tangent space to $S(t)\Sigma$ at $S(t)u_0$ determines uniquely an orthogonal projector $Q(t)$. We call the pair (u_0,Q_0) a contact element.

Proposition 15.1. The contact element $(u(t),Q(t))$ evolves acording to the equations

(15.12) $u(t) = S(t)u_0$

(15.13) $\frac{d}{dt} Q(t) + (I - Q(t))T(t)Q(t) + Q(t)T(t)^*(1 - Q(t)) = 0$

(15.14) $Q(0) = Q_0$

Here $T(t)$ depends on u_0 also and is the linearized operator

(15.15) $T(t)v = \varepsilon A^2 v + Av + B(S(t)u_0,v) + B(v,S(t)u_0)$

Proof. Let v_0 be a fixed element in H. Let $v_1(t),\ldots,v_N(t)$ be solutions of

(15.16) $\frac{dv}{dt} + T(t)v = 0$

with initial data v_i^0, $i = 1,\ldots,N$. Assume that Q_0 is the orthogonal projector in H onto the linear span of $v_1^0,\ldots,v_N^0$. Then $Q(t)$ will be the orthogonal projector onto the linear span of $v_1(t),\ldots,v_N(t)$. From (13.24) we know that

(15.17) $\frac{1}{2}\frac{d}{dt} |v_1\wedge\ldots\wedge v_N|^2 + Tr(TQ)|v_1 \wedge \ldots \wedge v_N|^2 = 0$

Let us consider the element of $\Lambda^{N+1}H$

$$v_0 \wedge v_1(t)\wedge \ldots \wedge v_N(t) = (I - Q(t))v_0 \wedge v_1(t)\wedge \ldots \wedge v_N(t).$$

Its time evolution is given by

$$\frac{d}{dt} (v_0\wedge v_1(t)\wedge\ldots\wedge v_N(t)) + [T(t)]_{N+1}(v_0\wedge\ldots\wedge v_N)$$
$$- T(t)v_0\wedge v_1(t) \wedge \ldots \wedge v_N(t) = 0$$

Therefore

$$\frac{1}{2}\frac{d}{dt} |v_0\wedge v_1\wedge\ldots\wedge v_N|^2 + Tr(T(t)\bar{Q}(t))|v_0 \wedge \ldots \wedge v_N|^2$$
$$- (T(t)v_0\wedge v_1 \wedge \ldots\wedge v_N;v_0 \wedge v_1 \wedge \ldots\wedge v_N) = 0$$

where $\tilde{Q}$ is the orthogonal projector on the span of $v_0, v_1(t), \ldots, v_N(t)$. (It was denoted $Q(v_0, v_1(t), \ldots, v_N(t))$ in section 13.) The notation T_{N+1} stands for $T_{N+1} = T \wedge 1 \wedge \ldots \wedge 1 \wedge + \ldots + 1 \wedge \ldots 1 \wedge T$ and was introduced in section 13.

We remark that

$$(Tv_0 \wedge \ldots \wedge v_N ; v_0 \wedge \ldots \wedge v_N) = ((1-Q)Tv_0 \wedge v_1 \wedge \ldots \wedge v_N ; v_0 \quad \ldots \quad v_N)$$

$$= ((1-Q)Tv_0, v_0)|v_1 \wedge \ldots \wedge v_N|^2$$

Now

$$(15.18) \qquad |(1 - Q)v_0|^2 = \frac{|v_0 \wedge v_1 \wedge \ldots \wedge v_N|^2}{|v_1 \wedge \ldots \wedge v_N|^2}$$

Differentiating both sides with respect to t we get

$$\frac{1}{2}\frac{d}{dt}|(1 - Q)v_0|^2 = -Tr(T\tilde{Q})|(1 - Q)v_0|^2 + ((1 - Q)Tv_0, v_0) + Tr(TQ)|(1 - Q)v_0|^2$$

$$= -(Tr\ T(\tilde{Q} - Q))|(1 - Q)v_0|^2 + ((1 - Q)Tv_0, v_0)$$

$$= -(T(1 - Q)v_0, (1 - Q)v_0) + ((1 - Q)Tv_0, v_0)$$

$$= ((1 - Q)TQv_0, v_0)$$

Since $v_0 \in H$ is time independent and arbitrary, it follows from the parallelogram identity that

$$\frac{d}{dt}(1 - Q(t)) = (1 - Q(t))T(t)Q(t) + Q(t)T(t)^*(1 - Q(t))$$

which establishes (15.13)

Now we are going to investigate the time evolution of the position of the range of $Q(t)$ relative to the fixed coordinate system given by the functions $w_1, \ldots, w_k, \ldots$ (eigenfunctions of A). Let Q be a N-dimensional orthogonal projector. We define the numbers $\Lambda(Q)$, $\lambda(Q)$ by

(15.19) $\Lambda(Q) = \max\{(Ag,g) \mid |g| = 1,\ Qg = g,\ g \in \mathcal{D}(A)\}$

(15.20) $\lambda(Q) = \min\{(Ag,g) \mid |g| = 1,\ Qg = 0,\ g \in \mathcal{D}(A)\}.$

From the minimax and maximin theorems it follows that

(15.21) $\Lambda(Q) \geq \lambda_N$

(15.22) $\lambda(Q) \leq \lambda_{N+1}$

where again, N is the dimension of Q and $\{\lambda_j\}$ is the sequence of repeated eigenvalues of A.

Suppose Q(t) evolves according to (15.13). Then the numbers $\Lambda(t) = \Lambda(Q(t))$ and $\lambda(t) = \lambda(Q(t))$ will satisfy certain differential inequalities that we wish to investigate. We will study $\Lambda(t)$, the study of $\lambda(t)$ being entirely similar. Let $t_0 > 0$ be fixed. Then $\Lambda(t_0)$ is actually the largest eigenvalue of the operator $Q(t_0)AQ(t_0)$. Assume that g is an eigenvector corresponding to it. Thus

(15.23) $Q(t_0)g = g,\quad |g| = 1,\ Q(t_0)Ag = \Lambda(t_0)g.$

The equality (15.23) is valid at $t = t_0$. We do not vary g. For t close to t_0 we consider the vector $\frac{Q(t)g}{|Q(t)g|}$. From the definition of $\Lambda(t)$ it follows that

$$\Lambda(t) \geq \frac{(AQ(t)g,Q(t)g)}{|Q(t)g|^2} = E(t).$$

At $t = t_0$ we have equality. Therefore for $t < t_0$

$$\frac{\Lambda(t) - \Lambda(t_0)}{t - t_0} \leq \frac{E(t) - E(t_0)}{t - t_0}.$$

Consequently, if we denote by $\overline{d}_\ell$ the derivative

$$\overline{d}_\ell(\Lambda(t))\Big|_{t=t_0} = \limsup_{\substack{t \to t_0 \\ t < t_0}} \frac{\Lambda(t) - \Lambda(t_0)}{t - t_0}$$

we obtain

$$\overline{d}_{\ell}(\Lambda(t))\Big|_{t=t_0} \leq \frac{d}{dt} E(t)\Big|_{t=t_0}$$

Now, it is easy to compute $\frac{d}{dt} E(t)|_{t=t_0}$. In order to make the computation easier let us note that

$$\frac{d}{dt} |Q(t)g|^2\Big|_{t=t_0} = 0. \tag{15.24}$$

Indeed, using (15.13), we have

$$\frac{d}{dt} |Q(t)g|^2 = 2(\frac{d}{dt} Q(t)g,g) = -2((1 - Q)TQg - QT^*(1 - Q)g,g)$$

$$= -4((1 - Q)g,TQg).$$

At $t = t_0$, $(1 - Q)g = 0$.

Differentiating $Q(t)$ and setting $t = t_0$ we get

$$\frac{d}{dt} E(t)\Big|_{t=t_0} = (A\dot{Q}g,g) + (Ag,\dot{Q}g) = 2(Ag,\dot{Q}g) = -2(Ag,(1 -Q)Tg).$$

We used $\dot{Q}$ for DQ/dt and (15.13). We obtained the inequality

$$\overline{d}_{\ell}(\Lambda(t))\Big|_{t=t_0} \leq -2(Ag,(1 - Q)Tg) \tag{15.25}$$

Now $Tg = \varepsilon A^2g + Ag + B(S(t_0)u_0,g) + B(g,S(t_0)u_0)$. Since $(1 - Q)$ is selfadjoint we have

$$-2(Ag,(1 - Q)Tg) = -2((1 - Q)Ag,Tg) = -2((A - \Lambda(t_0))g,Tg)$$

$$= -2\varepsilon((A - \Lambda)g,A^2g) - 2((A - \Lambda)g,Ag) - 2((A - \Lambda)g,B(u,g) + B(g,u)).$$

We denoted $S(t_0)u_0 = u$ and dropped the t_0 dependence in $\Lambda = \Lambda(t_0)$.

We compute the term $-2\varepsilon((A - \Lambda)g,A^2g)$ first. Let us remark that

$$((A - \Lambda)g,g) = 0$$

This follows from the fact that

$$\Lambda = (Ag,g), \quad |g| = 1.$$

This means that we can substitute A^2g by $(A^2 - c)g$ in $((A - \Lambda)g, A^2g)$ for any constant c. We take this constant to be Λ^2:

$$-2\varepsilon((A - \Lambda)g,\ A^2g) = -2\varepsilon((A - \Lambda)g,\ (A^2 - \Lambda^2)g).$$

Now $A^2 - \Lambda^2 = (A + \Lambda)(A - \Lambda)$. Therefore we obtain

$$(15.26) \qquad -2\varepsilon((A - \Lambda)g, A^2g) \leq -2\varepsilon\Lambda|(A - \Lambda)g|^2.$$

On the other hand, we estimate the term

$$-2((A - \Lambda)g,\ B(u,g) + B(g,u))$$

as follows

$$2|((A - \Lambda)g,\ B(u,g) + B(g,u))| \leq 2|(A - \Lambda)g||B(g,u)| + 2|(A - \Lambda)g||B(u,g)|$$

$$\leq 2|(A - \Lambda)g|[|g|_{L^\infty}\|u\| + |u|_{L^\infty}\|g\|].$$

Now $\|g\| = \Lambda^{1/2}$ since $\|g\|^2 = (Ag,g) = \Lambda$. Moreover, $|g|_{L^\infty} \leq c\|g\|(1 + \log_+|Ag|)$ (see 15.10) and since $Ag = (A - \Lambda)g + \Lambda g$ we get an estimate of the type

$$|g|_{L^\infty} \leq c\Lambda^{1/2}[1 + \log_+(|(A - \Lambda)g| + \Lambda)]$$

On the other hand, the estimates (15.11), (15.8) for $u(t) = S(t)u_0$ are of the form

$$(15.27) \qquad \|u\| \leq k_1$$

$$(15.28) \qquad \|u\|_{L^\infty} \leq k_2\log_+(1 + |Au_0|)$$

with k_1, k_2 constants depending on $\|u_0\|$, ε and $|f|$ only. Combining all these together we get

$$2|(A - \Lambda)y|[|y|_\infty \|u\| + |u|_\infty \|g\|]$$

$$\le c\Lambda^{1/2}|(A - \Lambda)y|[k_1(1 + \log_+(\Lambda + |(A - \Lambda)y|)) + k_2\log_+(1 + |Au_0|)]$$

Let us denote $|(A - \Lambda)y| = \delta$. Then

$$(15.29) \quad \bar{d}_\ell\Lambda \le -2\varepsilon\Lambda\delta^2 - 2\delta^2 + c\Lambda^{1/2}\delta[k_1(1+\log_+(\Lambda+\delta)) + k_2\log_+(1 + |Au_0|)]$$

Now we note that $\Lambda \ge 1$ because of the Poincaré inequality. Therefore

$$\log(\Lambda + \delta) \le \log(\Lambda + \delta\Lambda) = \log \Lambda + \log(1 + \delta)$$

$$\le \log \Lambda + \log(\delta\sqrt{\Lambda} + 1) \le \log \Lambda + (\delta\sqrt{\Lambda})^{1/2}$$

(We used the inequality $\log(1 + x) \le \sqrt{x}$, valid for $x \ge 0$.) Then

$$ck_1\Lambda^{1/2}\delta(1 + \log(\Lambda + \delta)) \le (ck_1\log \Lambda)\Lambda^{1/2}\delta + ck_1\Lambda^{1/2}\delta + (ck_1)(\Lambda^{1/2}\delta)^{3/2}$$

$$\le \frac{\varepsilon}{2}\Lambda\delta^2 + c_\varepsilon k_1^2(\log \Lambda)^2 + c_\varepsilon(k_1^4 + k_1^2).$$

From (15.29) we obtain

$$(15.30) \quad \bar{d}_\ell(\Lambda) \le -(\varepsilon\Lambda + 2)\delta^2 + c_\varepsilon[k_1^2(1+(\log \Lambda)^2) + k_1^4 + k_2^2(1+\log(|Au_0|^2+1))^2]$$

Now $\delta^2 = |(A - \Lambda)y|^2 \ge \text{dist}(\Lambda,\sigma(A))^2$ where $\sigma(A) = \{\lambda_1,\lambda_2,\ldots,\lambda_j,\ldots\}$. Indeed, since $|y| = 1$ we have

$$|(A - \Lambda)y|^2 = \sum_{j=1}^{\infty}(\lambda_j - \Lambda)^2 y_j^2 \ge \text{dist}(\Lambda,\sigma(A))^2.$$

We proved therefore

Proposition 15.2. Let $u_0 \in \mathcal{D}(A)$. Let $S(t)u_0 = u(t)$ be a solution of (15.3), (15.4). Let Q_0 be an arbitrary N dimensional projector. Let $Q(t)$ evolve according to (15.13), (15.14). Let $\Lambda(t) = \Lambda(Q(t))$, $\lambda(t) = \lambda(Q(t))$ where $\Lambda(Q)$, $\lambda(Q)$ are defined in (15.19), (15.20). Denote by

$$\delta(\Lambda) = \mathrm{dist}(\Lambda, \sigma(A)) = \min_{j=1,2,\ldots} |\Lambda - \lambda_j|.$$

There exist constants k_3, k_4 depending on $\|u_0\|$, ε, $|f|$ only such that

(15.31) $$\overline{d}_\ell(\Lambda) \le -\Lambda\varepsilon(\delta(\Lambda))^2 + k_3[\log \Lambda + \log(1 + |Au_0|^2)]^2 + k_4$$

(15.32) $$\underline{d}_\ell(\lambda) \ge \varepsilon\lambda(\delta(\lambda))^2 - k_3[\log \lambda + \log(1 + |Au_0|^2)]^2 - k_4$$

Theorem 15.3 (Spectral Blocking Property). Let $u_0 \in \mathcal{D}(A)$. Assume

(15.33) $$\|u_0\| \le R_0$$

(15.34) $$|Au_0| \le R$$

There exist constants $k_3(R_0, \varepsilon, |f|)$, $k_4(R_0, \varepsilon, |f|)$ (independent of R) such that if λ_m satisfies the gap condition

(15.35) $$\varepsilon \frac{\lambda_m + \lambda_{m+1}}{2} (\lambda_{m+1} - \lambda_m)^2 > k_4 + k_3[\log \frac{\lambda_m + \lambda_{m+1}}{2} + \log(1 + R^2)]^2$$

then

(a) If Q_0 is a projector of dimension $N \le m$ such that

$$\Lambda(Q_0) \le \frac{\lambda_m + \lambda_{m+1}}{2}$$

then

$$\Lambda(Q(t)) \le \frac{\lambda_m + \lambda_{m+1}}{2}$$

for all $t \ge 0$.

(b) If Q_0 is a projector of dimension $N \geq m$ such that

$$\lambda(Q_0) \geq \frac{\lambda_m + \lambda_{m+1}}{2}$$

then

$$\lambda(Q(t)) \geq \frac{\lambda_m + \lambda_{m+1}}{2}$$

for all $t \geq 0$.

Let us note that the gap condition (15.35) can e easily fulfilled, for infinitely many λ_m's. Indeed, the λ_m's are sums of squares and as such they have discrete gaps and behave like multiples of the index m. Actually a much stronger property holds ([R1])

(15.36) $\lambda_{m+1} - \lambda_m \geq c \log \lambda_m$

for infinitely many m's. We do not use this property for the present equation. All we use is $\lambda_{m+1} - \lambda_m \geq \delta > 0$ for infinitely many m's and $\frac{\lambda_m}{(\log \lambda_m)^2} \to \infty$, that is $\lambda_m \to \infty$. Of course, implicit in the proof is the fact that A^2 has large enough gaps to dominate the nonlinearity. However, if the nonlinearity would have been of reaction-diffusion type (no derivatives) then the gap condition requirement would have been much weakened.

The proof of the theorem is immediate form (15.31) and (15.32) because they imply that the point $\frac{\lambda_m + \lambda_{m+1}}{2}$ is a repelling point for the evolution of $\Lambda(t)$ from left to right and for the evolution of $\lambda(t)$ from right to left.

In the construction of the inertial manifold we will have to let R to depend on λ_m: $R^2 = \lambda_m R_0^2$. Since the constants k_2, k_3 depend on R_0 and $|f|$ only, condition (15.35) with $R = \lambda_m R_0^2$ creates no difficulty.

Theorem 15.3 has important consequences. First let us discuss the consequence of point (a) of the theorem for N = 1. Let v(t) be a solution of

$$\frac{dv}{dt} + \varepsilon A^2 v + Av + B(S(t)u_0, v) + B(v, S(t)u_0) = 0 \tag{15.37}$$

$$v(0) = v_0 \tag{15.38}$$

where u_0 satisfies $\|u_0\| \le R_0$, $|Au_0| \le R$. Then the projection on the direction $v(t)$ constitutes $Q(t)$ for $N = 1$. The quantity

$$\Lambda(Q(t)) = \frac{(Av(t), v(t))}{|v(t)|^2} = \frac{\|v\|^2}{|v|^2} .$$

Let us assume that $(\lambda_m, \lambda_{m+1})$ is any of the gaps satisfying (15.35). Point (a) of Theorem 15.3 is an invariance statement about the locally compact cone in H

$$K_m = \{w \in V \mid \|w\|^2 \le \frac{\lambda_m + \lambda_{m+1}}{2} |w|^2\}. \tag{15.39}$$

If $v(t_0)$ belongs to K_m then $v(t)$ belongs to K_m for all $t \ge t_0$. Now, there are two possibilities: either $v(t)$ stays outside K_m for all $t \ge 0$ or it enters K_m in finite time, never to leave it again. We will show that the first alternative implies very strong exponential decay of $|v(t)|$. In other words, K_m are "slow cones" and their complements "fast cones". Indeed, from the first energy equation for $v(t)$:

$$\frac{1}{2}\frac{d}{dt}|v|^2 + \|v\|^2 \le c|v|\,\|v\|\,\|u\| \le \frac{\|v\|^2}{2} + c|v|^2\|u\|^2.$$

We used (6.20) with $s_2 = s_3 = 1/2$, $s_2 = 0$. Now from (15.8), $\|u\|^2 \le R_0^2 + |f|^2$. We infer

$$\frac{d}{dt}|v|^2 + \left(\frac{\|v\|^2}{|v|^2} - c(R_0^2 + |f|^2)\right)|v|^2 \le 0$$

Therefore, if

$$\frac{\lambda_m + \lambda_{m+1}}{2} > 2c(R_0^2 + |f|^2) \tag{15.40}$$

then, as long as $\frac{\|v\|^2}{|v|^2} > \frac{\lambda_m + \lambda_{m+1}}{2}$ we have

$$(15.41) \qquad |v(t)|^2 \leq |v(0)|^2 \exp\left(- \frac{\lambda_m + \lambda_{m+1}}{4} t\right)$$

The condition (15.40) is a only largeness condition on m and it can be achieved simultaneously with (15.35) even in the case $R^2 = \lambda_m R_0^2$.

The property just described is called "strong squeezing" in [CFNT]. We have the similar property for differences $S(t)u_0 - S(t)u_1$ if both $\|u_0\| \leq R_0$, $\|u_1\| \leq R_0$, $|Au_0| \leq R$, $|Au_1| \leq R$.

Theorem 15.4 (Invariance of the slow cones K_m). Let $u_0, u_1 \in \mathcal{D}(A)$. Assume $\|u_0\| \leq R_0$, $\|u_1\| \leq R_0$, $|Au_0| \leq R$, $|Au_1| \leq R$. Let v(t) denote either the solution of the linearized equation (15.37) or the difference $S(t)u_0 - S(t)u_1$. Let $(\lambda_m, \lambda_{m+1})$ be a gap satisfying (15.35) and assume that the gap is located far enough so that (15.40) be valid. Then, if $v(t) \in K_m$ then $v(s) \in K_m$, for all $s \geq t$. Moreover, either

(a) $|v(t)|^2 \leq |v(0)|^2 \exp\left(- \frac{\lambda_m + \lambda_{m+1}}{4} t\right)$ for all $t \geq 0$

or

(b) There exists $0 \leq t_0 < \infty$ such that the inequality in (a) is valid for $t \leq t_0$ and, for $t \geq t_0$, $v(t) \in K_m$.

Let us consider now a different type of cones

$$(15.42) \qquad C_{m,\gamma} = \{v \in H \mid |(1 - P_m)v|^2 \leq \gamma^2 |P_m v|^2\}$$

where $\gamma > 0$ and P_m is the projection onto the linear span of the first m eigenfunctions of A. Let v(t) be a solution of (15.37) where $\|u_0\| \leq R_0$, $|Au_0| \leq R$. Let us denote by $q(t) = (1 - P_m)v(t)$ and by $p(t) = P_m v(t)$.

Then

$$(15.43) \quad \frac{1}{2}\frac{d}{dt}[|q|^2 - \gamma^2|p|^2] + \varepsilon|Aq|^2 - \gamma^2\varepsilon|Ap|^2$$

$$+ \|q\|^2 - \gamma^2\|p\|^2 + (B(u,v) + B(v,u),\ q-\gamma^2 p) = 0$$

Now

$$|(B(u,v) + B(v,u), q-\gamma^2 p)| \le c_\gamma \|u\|[\|q\| + \|p\|][|q| + |p|]$$

$$\le c_\gamma (R_0^2 + |f|^2)^{1/2} [\|q\| + \|p\|][|q| + |p|]$$

$$\le \frac{\varepsilon}{2}\|q\|^2 + \frac{1}{2\varepsilon} c_\gamma^2 (R_0^2 + |f|^2)(|q|^2 + |p|^2) + c_\gamma (R_0^2 + |f|^2)^{1/2} \|p\| [|q| + |p|]$$

$$\le \frac{\varepsilon}{2}\|q\|^2 + \frac{1}{2\varepsilon} c_\gamma^2 (R_0^2 + |f|^2)(|q|^2 + |p|^2) + c_\gamma \lambda_m^{1/2} (R_0^2 + |f|^2)[|q| + |p|].$$

Therefore computing $\frac{1}{2}\frac{d}{dt}[|q|^2 - \gamma^2|p|^2]$ at a point where $|q| = \gamma|p|$ yields

$$\frac{1}{2}\frac{d}{dt}[|q|^2 - \gamma^2|p|^2]_{|q|=\gamma|p|} \le -\varepsilon|Aq|^2 + \gamma^2\varepsilon|Ap|^2 + \frac{\varepsilon}{2}\|q\|^2$$

$$+ k_5|p|^2 + k_6 \lambda_m^{1/2}|p| + \rho^2\|p\|^2 - \|q\|^2.$$

Here k_5 and k_6 depend on R_0, $|f|$, ε and γ alone. Now

$$-\varepsilon|Aq|^2 + \frac{\varepsilon}{2}\|q\|^2 + \gamma^2(\varepsilon Ap|^2 + \|p\|^2) - \|q\|^2$$

$$\le (-\varepsilon\lambda_{m+1} - 1 + \tfrac{\varepsilon}{2})\|q\|^2 + \gamma^2(\varepsilon\lambda_m^2 + \lambda_m)|p|^2 \le$$

$$\le (-\varepsilon\lambda_{m+1} - 1 + \tfrac{\varepsilon}{2})\lambda_{m+1}|q|^2 + \gamma^2(\varepsilon\lambda_m^2 + \lambda_m)|p|^2$$

$$\le [-\varepsilon\lambda_{m+1}^2 + \varepsilon\lambda_m^2 + \frac{\varepsilon}{2}\lambda_{m+1}]\gamma^2|p|^2$$

Thus

$$\frac{1}{2}\frac{d}{dt}[|q|^2 - \gamma^2|p|^2]\Big|_{|q|=\gamma|p|} \le |p|^2[-\varepsilon\lambda_{m+1}^2 + \varepsilon\lambda_m^2 + \frac{\varepsilon}{2}\lambda_{m+1} + \tilde{k}_7 + \frac{\varepsilon}{2}\lambda_m]$$

where $\tilde{k}_7 = \frac{1}{\gamma^2}(k_5 + \frac{k_6^2}{2\varepsilon})$. Therefore

$$(15.44) \qquad \frac{1}{2}\frac{d}{dt}[|q|^2 - \gamma^2|p|^2]\Big|_{|q|=\gamma|p|} < 0$$

provided

$$(15.45) \qquad \varepsilon(\lambda_{m+1} - \lambda_m)(\lambda_{m+1} + \lambda_m) \ge \frac{\varepsilon}{2}(\lambda_{m+1} + \lambda_m) + \tilde{k}_7.$$

This condition can be fulfilled easily, for instance, if $\lambda_{m+1} - \lambda_m \ge 1$ (which in our case requires only $\lambda_{m+1} \ne \lambda_m$ since λ_j are integers) and if

$$(15.46) \qquad \frac{\lambda_{m+1} + \lambda_m}{2} \ge k_7 = \frac{\tilde{k}_7}{\varepsilon}.$$

We have therefore

Theorem 15.5 (Invariance of the slow cones $C_{m,\gamma}$). Let u_0, u_1 be in $\mathcal{D}(A)$. Assume $\|u_0\| \le R_0$, $\|u_1\| \le R_0$. Let $v(t)$ denote either the solution of the linearized equation (15.37) or the difference $S(t)u_0 - S(t)u_1$. There exists $k_7 = k_7(R_0, |f|, \gamma, \varepsilon)$ such that if $\lambda_{m+1} > \lambda_m$ and $\frac{\lambda_{m+1} + \lambda_m}{2} \ge k_7$ then the cones $C_{m,\gamma}$ are invariant: if $v(t_0) \in C_{m,\gamma}$ then $v(t) \in C_{m,\gamma}$ for all $t \ge t_0$.

We have at this stage all the ingredients for the construction of inertial manifolds for the equation (15.1). Let us check some of the consequences of Theorems (15.3)-(15.5) for the universal attractor X. First let us estimate the dimension of the attractor and the decay of volume elements. Let $\varphi_1,\ldots,\varphi_N$ be orthonormal in H. Let Q be the linear space spanned by $\varphi_1,\ldots,\varphi_N$.

Let T be the linearized (15.15)

$$Tv = \varepsilon A^2 v + Av + B(u,v) + B(v,u).$$

Then $$\mathrm{Tr}(TQ) \geq \varepsilon \sum_{i=1}^{N} (A^2 \varphi_i, \varphi_i) + \sum_{i=1}^{N} B(\varphi_i, u, \varphi_i).$$

Thus $$\mathrm{Tr}(TQ) \geq \varepsilon \sum_{i=1}^{N} |A\varphi_i|^2 - c\|u\| \sum_{i=1}^{N} \|\varphi_i\|.$$

Since $\|\varphi_i\| \leq |\varphi_i|^{1/2}|A\varphi_i|^{1/2} = |A\varphi_i|^{1/2}$ it follows that

$$\sum_{i=1}^{N} \|\varphi_i\| \leq \sum_{i=1}^{N} |A\varphi_i|^{1/2} \leq N^{3/4}\Big(\sum_{i=1}^{N} |A\varphi_i|^2\Big)^{1/4}$$

Thus

$$\mathrm{Tr}(TQ) \geq \frac{\varepsilon}{2} \sum_{i=1}^{N} |A\varphi_i|^2 - c_\varepsilon N \|u\|^{4/3}.$$

If $u = S(t)u_0$, $\|u_0\| \leq R_0$ then $\|u\|^2 \leq R_0^2 e^{-t} + |f|^2$ (see 15.8). On the other hand, because φ_i are orthonormal

$$\sum_{i=1}^{N} |A\varphi_i|^2 \geq \lambda_1^2 + \cdots + \lambda_N^2 .$$

Therefore

$$\frac{1}{t}\int_0^t \mathrm{Tr}\, T(s)Q(s)ds \geq \frac{\varepsilon}{2}(\lambda_1^2 + \cdots \lambda_N^2) - cN|f|^{4/3} \tag{15.47}$$

if $t \geq t_0(R_0)$.

This estimate implies (see section 14)

$$\mu_1 + \cdots \mu_N \leq -\frac{\varepsilon}{2}(\lambda_1^2 + \cdots + \lambda_N^2) + cN|f|^{4/3} \tag{15.48}$$

for all N = 1,2,... . Now since $\lambda_j \geq c_0 j$, the right-hand side of (15.48) becomes negative for

(15.49) $N \geq c_\varepsilon |f|^{2/3}$.

Therefore we have

Theorem 15.6. Let X be the universal attractor of equation (15.3).

(15.50) $d_H(X) \leq d_M(X) \leq c|f|^{2/3}$

Now let us fix R_0 large enough so that

(15.51) $2 \sup_{u \in X} \|u\| \leq R_0$

From (15.8) it follows that $R_0 = 2|f|$ would do.

Let $(\lambda_m, \lambda_{m+1})$ be a gap (i.e., $\lambda_{m+1} > \lambda_m$) and assume that (15.46) is satisfied for the constant k_7 determined by R_0, $|f|$ and $\gamma = 1/3$ (we choose $\gamma = 1/3$ rather arbitrarily). Assume also that $N = m-1$ satisfies (15.49). From the assumption (15.46) it follows that for any two $u_1, u_2 \in X$, $u_1 - u_2 \in C_{m,1/3}$. This is a direct consequence of the fact that outside $C_{m,1/3}$ there is exponential decay for differences $S(t)u_1^0 - S(t)u_2^0$, just as in the case of the cones K_m.

Proposition 15.7. Let $\gamma = 1/3$ (for the sake of being specific). Let $\lambda_m < \lambda_{m+1}$, and assume λ_m satisfy (15.46). Then the set

(15.52) $C_{m,X} = \{u | u - x \in C_{m,1/3} \text{ for all } x \in X\}$

is invariant

(15.53) $S(t)(C_{m,X} \cap B^V_{R_0}) \subset C_{m,X}$.

Moreover, for arbitrary $u_0 \in B^V_{R_0}$ either

(a) $\text{dist}(S(t)u_0, X) \leq \max_{u \in X} |u - u_0| \exp(-\frac{\lambda_m + \lambda_{m+1}}{4} t)$ for all t or

(b) the inequality in (a) is valid for $t \le t_0 < \infty$ and for $t \ge t_0$, $S(t)u_0 \in C_{m,\chi}$.

We are going to consider the set $\Sigma = \bigcup_{t>0} S(t)\Gamma$ where Γ will be a simple N-1 dimensional smooth compact surface in the N dimensional space $H_N = P_N H$. Our aim is to show that $\overline{\Sigma} = Y$ is an inertial manifold.

Definition 15.8. A set $Y \subset H$ is an inertial manifold for the equation (15.3) if

(i) $S(t)Y \subset Y$ for all $t \ge 0$

(ii) Y is a finite dimensional, Lipschitz manifold

(iii) There exists a constant c_Y such that for every $u_0 \in H$ there exists C, t_0 depending on $|u_0|$ such that

$$\text{dist}(S(t)u_0, Y) \le C\exp(-c_Y t) \quad \text{for } t \ge t_0. \tag{15.54}$$

Actually the inertial manifold that we construct will satisfy several other important properties. Let us prepare first the initial data set Γ. We take Γ to be an ellipsoid in H_m.

$$\Gamma = \{u \in V \mid P_m u = u,\ \|u\| = R_0\} \tag{15.55}$$

The number R_0 can be chosen

$$R_0 = 4\|f\| \tag{15.56}$$

The number m is going to be determined below.

Let $u \in \Gamma$. We consider the vector

$$N(u) = \epsilon A^2 u + Au + B(u,u) - f \tag{15.57}$$

We consider the linear space $T_u(\Gamma)$, tangent to Γ in $H_m = P_m H$ at u. We take the direct sum $T_u(\Gamma) \oplus N(u)\mathbb{R}$. This is the initial tangent space at our integral manifold. Let us denote by $Q(u)$ the orthogonal projector in

H from H onto $T_u(\Gamma) \oplus N(u)\mathbb{R}$. The requirements that we wish to impose on Γ are

(I) $\Lambda(Q(u)) < \dfrac{\lambda_m + \lambda_{m+1}}{2}$ for all $u \in \Gamma$

(II) $\lambda(Q(u)) > \dfrac{\lambda_m + \lambda_{m+1}}{2}$ for all $u \in \Gamma$

(III) $(N(u), \nu(u)) > 0$ for all $u \in \Gamma$, where $\nu(u)$ is the external normal to Γ in $P_m H$ at u.

(IV) $\Gamma \subset C_{m,X}$

(V) $T_u(\Gamma) \oplus N(u)\mathbb{R} \subset C_{m,X}$

The conditions (I) and (II) will insure that the integral manifold $\Sigma = \bigcup_{t > 0} S(t)\Gamma$ is "spectrally blocked". Condition (III) is a coercivity condition. We check first what kind of restrictions do conditions (I)-(V) impose on m.

We start by showing that in our case condition (III) is automatically fulfilled without restriction for m. Indeed, the normal $\nu(u)$ can be computed in our case

(15.58) $\nu(u) = A \dfrac{u}{|u|}$.

Indeed, $\|u\| = R_0$ is an ellipsoid in H_m:

$$\sum_{j=1}^{m} \lambda_j u_j^2 = R_0^2 , \qquad (u_j = (u,w_j)).$$

Now, in view of the identity

$$B(u,u,Au) = 0$$

we have

(15.59) $(N(u), \nu(u)) \geq \dfrac{\varepsilon|A^{3/2}u|^2 + \frac{1}{2}|Au|^2}{|u|} > 0$ for all $u \in \Gamma$.

We used here $|Au| \geq 2|f|$ for $u \in \Gamma$. In order to check (I) and (II) we denote by $\Lambda_1' \le \Lambda_2' \le \cdots \Lambda_m'$ the eigenvalues of $Q(u)AQ(u)$. In particular, $\Lambda_m' = \Lambda(Q(u))$. We know $\Lambda(u) \ge \lambda_m$ so

$$0 \le \Lambda(u) - \lambda_m = \Lambda_1' + \cdots + \Lambda_m' - \lambda_1 - \lambda_2 - \cdots - \lambda_m$$

$$- (\Lambda_1' + \cdots + \Lambda_{m-1}' - \lambda_1 - \cdots \lambda_{m-1})$$

$$\le \Lambda_1' + \cdots \Lambda_m' - (\lambda_1 + \cdots \lambda_m)$$

because

$$\Lambda_1' + \cdots \Lambda_{m-1}' \ge \lambda_1 + \cdots + \lambda_{m-1}.$$

Thus

(15.60) $$0 \le \Lambda(u) - \lambda_m \le \operatorname{Tr} Q(u)AQ(u) - \operatorname{Tr} P_m A P_m .$$

Similarly

(15.61) $$0 \le \lambda_{m+1} - \lambda(u) \le \operatorname{Tr} Q(u)AQ(u) - \operatorname{Tr} P_m A P_m$$

Now we decompose $Q(u)$:

(15.62) $$Q(u) = \omega \otimes \omega + Q_\tau(u)$$

where Q_τ is the orthogonal projector from H onto the tangent space $T_u(\Gamma)$ at Γ to u. The vector ω is the normalized component of $N(u)$ orthogonal to $T_u(\Gamma)$:

(15.63) $$\omega = \frac{(1 - Q_\tau)N(u)}{|(1 - Q_\tau)N(u)|} .$$

The notation $\omega \otimes \omega$ stands for the one dimensional projector in the direction ω

$$v \longmapsto (\omega, v)\,\omega .$$

On the other hand, obviously

$$P_m = \nu(u) \otimes \nu(u) + Q_\tau. \tag{15.64}$$

Then

$$\text{Tr } Q(u)AQ(u) - \text{Tr } P_m A P_m = \|\omega\|^2 - \|\nu(u)\|^2.$$

Now, from (15.64), $1 - Q_\tau = 1 - P_m + \nu(u) \otimes \nu(u)$. Since $\nu(u) \in P_m H$ it follows that

$$|(1 - Q_\tau)N(u)|^2 = |(1 - P_m)N(u)|^2 + (N(u),\nu(u))^2 \tag{15.65}$$

$$\|(1 - Q_\tau)N(u))\|^2 = \|(1 - P_m)N(u)\|^2 + (N(u),\nu(u))^2\|\nu(u)\|^2 \tag{15.66}$$

In the computation of $\|\omega\|^2 - \|\nu(u)\|^2$ the cross terms $(N(u),\nu(u))^2\|\nu(u)\|^2$ cancel each other:

$$\|\omega\|^2 - \|\nu(u)\|^2 = \frac{\|(1 - P_m)N(u)\|^2 - \|\nu(u)\|^2|(1 - P_m)N(u)|^2}{|(1 - P_m)N(u)|^2 + (N(u),\nu(u))^2}.$$

Thus

$$\text{Tr } Q(u)AQ(u) - \text{Tr } P_m A P_m \leq \frac{\|(1 - P_m)N(u)\|^2}{(N(u),\nu(u))^2} \tag{15.67}$$

The conditions (I) and (II) will be fulfilled if

$$\frac{\|(1 - P_m)N(u)\|^2}{(N(u),\nu(u))^2} < \frac{\lambda_{m+1} - \lambda_m}{2} \tag{15.68}$$

Now the quantity

$$\begin{aligned}(1 - P_m)N(u) &= (1 - P_m)\varepsilon A^2 u + (1 - P_m)(B(u,u) - f) + (1 - P_m)Au \\ &= (1 - P_m)(B(u,u) - f)\end{aligned}$$

since $u \in P_m H$. So

$$\|(1-P_m)N(u)\|^2 = \|(1-P_m)(B(u,u)-f)\|^2$$

$$\leq 2|A^{1/2}(1-P_m)B(u,u)|^2 + 2\|f\|^2.$$

Now

$$|A^{1/2}(1-P_m)B(u,u)|^2 \leq \frac{1}{\lambda_{m+1}}|A(1-P_m)B(u,u)|^2 \leq \frac{1}{\lambda_{m+1}}|AB(u,u)|^2.$$

Using the inequality $|AB(u,u)| \leq c|Au||A^{3/2}u|$ (see (10.8) with s = 2) we yet

$$\|(1-P_m)N(u)\|^2 \leq \frac{c}{\lambda_{m+1}}|A^{3/2}u|^4 + \|f\|^2$$

On the other hand, from (15.59)

$$(N(u),\nu(u))^2 \geq \mathrm{Max}\{\frac{\varepsilon}{|u|^2}|A^{3/2}u|^4, \frac{1}{4}\frac{|Au|^4}{|u|^2}\}.$$

Thus

$$\frac{\|(1-P_m)N(u)\|^2}{(N(u),\nu(u))^2} \leq \frac{c}{\lambda_{m+1}}|u|^2 + \frac{4\|f\|^2|u|^2}{|Au|^4} \leq \frac{cR_0^2}{\lambda_{m+1}} + \frac{4\|f\|^2}{R_0^2} = \frac{16c\|f\|^2}{\lambda_{m+1}} + \frac{1}{4}.$$

Since if $\lambda_{m+1} - \lambda_m > 0$ we have $\lambda_{m+1} - \lambda_m \geq 1$, condition (15.68) will be fulfilled if λ_{m+1} is large enough:

$$(15.69) \qquad \lambda_{m+1} \geq c\|f\|^2.$$

We see thus that conditions I, II, III can be fulfilled if $(\lambda_m, \lambda_{m+1})$ is a gap, situated far enough. No requirement of largeness on the gap is needed. Now (IV) and (V) are easier to realize than (I) and (II).

Let us check IV: First let us note that from (15.8) it follows that

$$(15.70) \qquad \sup_{x\in X}\|x\| \leq |f|.$$

Take $u \in \Gamma$, $x \in X$. We need

$$|(1-P_m)(u-x)| \leq \frac{1}{3}|P_m(u-x)|.$$

Now $(1 - P_m)(u - x) = -(1 - P_m)x$ since $u \in P_m H$. Thus

$$|(1 - P_m)(u - x)| = |(1 - P_m)x| \leq \lambda_{m+1}^{-1/2} \|x\| \leq \lambda_{m+1}^{-1/2}|f|.$$

On the other hand

$$\frac{1}{3}|P_m(u - x)| \geq \lambda_m^{-1/2} \frac{1}{3} \|P_m(u - x)\| \geq \lambda_m^{-1/2} \frac{1}{3} [\|P_m u\| - \|P_m x\|]$$

$$= \lambda_m^{-1/2} \frac{1}{3} [\|u\| - \|P_m x\|] \geq \lambda_m^{-1/2} \frac{1}{3} \|u\| - \frac{\lambda_m^{-1/2}}{3} |f|.$$

Thus $u - x \in C_{m,1/3}$ provided

$$\left(\lambda_{m+1}^{-1/2} + \frac{\lambda_m^{-1/2}}{3}\right)|f| \leq \frac{1}{3} \lambda_m^{-1/2} R_0.$$

Multiplying by $\lambda_m^{1/2}$ and using $\lambda_m \leq \lambda_{m+1}$ we see that $u - x \in C_{m,1/3}$ provided

$$\frac{4}{3} |f| \leq \frac{1}{3} R_0 = \frac{4}{3} \|f\|.$$

No extra requirement is needed.

Finally, let us check (V). Let $v \in N(u)\mathbb{R} + T_u(\Gamma)$. Thus $v = \alpha\,\omega + \beta\tau$ with $\tau \in T_u(\Gamma)$, ω defined in (15.63) and α, β real numbers. $(1 - P_m)v = \alpha(1 - P_m)\omega$; $P_m v = \alpha P_m \omega + \beta\tau$. Since $|P_m v|^2 = \alpha^2 |P_m \omega|^2 + \beta^2 |\tau|^2 \geq \alpha^2 |P_m \omega|^2$ it is enough to check that

$$|(1 - P_m)\omega| \leq \frac{1}{3} |P_m \omega|.$$

Now we use the decomposition (15.64) and (15.65)

$$|(1 - P_m)\omega|^2 = \frac{|(1 - P_m)N(u)|^2}{|(1 - P_m)N(u)|^2 + (N(u), \nu(u))^2}$$

$$\frac{1}{9} |P_m \omega|^2 = \frac{1}{9} \frac{(N(u), \nu(u))^2}{|(1 - P_m)N(u)|^2 + (N(u), \nu(u))^2}.$$

Therefore what we need is

$$(15.71) \qquad |(1 - P_m)N(u)| \le \frac{1}{3} (N(u),\nu(u))$$

But condition (15.68) will imply (15.70) if

$$(15.72) \qquad \frac{\lambda_{m+1} - \lambda_m}{2\lambda_{m+1}} \le \frac{1}{9} .$$

This again is a mild requirement:

$$(15.73) \qquad \lambda_{m+1} \ge \frac{9(\lambda_{m+1} - \lambda_m)}{2} .$$

Summarizing we proved

Lemma 15.8. Let m be such that $\lambda_{m+1} > \lambda_m$ and λ_{m+1} is large enough:

$$(15.74) \qquad \lambda_{m+1} \ge \max(9 \frac{(\lambda_{m+1} - \lambda_m)}{2} , c_\varepsilon \|f\|^2)$$

Then the flat m-1 dimensional ellipsoid

$$(15.75) \qquad \Gamma = \{u \mid P_m u = u, \|u\| = R_0\}$$

with

$$(15.76) \qquad R_0 = 4\|f\|$$

satisfies properties (I)-(V).

Let us fix $R_0 = 4\|f\|$. Let us seek $m \ge 1$ satisfying (15.35) with $R^2 = \lambda_m R_0^2$, i.e.,

$$(15.77) \quad \varepsilon \frac{\lambda_{m+1} + \lambda_m}{2} (\lambda_{m+1} - \lambda_m)^2 > k_4 + k_3[\log \frac{\lambda_m + \lambda_{m+1}}{2} + \log(1+\lambda_m R_0^2)]^2$$

and (15.40), (15.46), (15.74). Assume also that m is large enough so that $N = m-1$ satisfies (15.49). The conditions (5.40), (15.46), (15.74) are all of the form $\frac{\lambda_{m+1} + \lambda_m}{2} \ge k$ where $k = k(R_0, \varepsilon, |f|)$. Since R_0, ε are fixed, $k = k(\|f\|)$ depends on $\|f\|$ alone. Condition (15.77) is of the type

$$\varepsilon\, \frac{\lambda_{m+1} + \lambda_m}{2} > k_9 + k_8\left(\log\left(\frac{\lambda_m + \lambda_{m+1}}{2}\right)\right)^2$$

because $\lambda_{m+1} - \lambda_m \geq 1$ if $\lambda_{m+1} > \lambda_m$. Condition (15.49) is expressed in terms of m,

$$m - 1 \geq c|f|^{2/3}.$$

Since $\lambda_m \geq c_0 m$ it follows that we can fulfill simultaneously conditions (15.40), (15.46), (15.74), (15.77) provided

(15.78) $\qquad \lambda_{m+1} > \lambda_m$

(15.79) $\qquad \dfrac{\lambda_{m+1} + \lambda_m}{2} \geq c_\varepsilon(\|f\|)$

with $c_\varepsilon(\|f\|)$ a sufficiently large positive constant. We fix now m such that $(\lambda_m, \lambda_{m+1})$ satisfy (15.78) and (15.79). Thus the spectral blocking property (Theorem 15.4) the $C_{m,1/3}$ cone invariance (Theorem 15.5) are valid. Moreover, surfaces of dimension m-1 decay exponentially ((15.48), (15.49)) and the initial surface $\Gamma = \{u \in P_m H \mid \|u\| = R_0\}$ has the properties (I)-(V).

We start studying the m dimensional surface $\Sigma = \bigcup_{t > 0} S(t)\Gamma$. Let us consider the maps s and σ given by

$$s:(0,\infty) \times \Gamma \to \Sigma \subset H, \quad s(t,u_0) = S(t)u_0$$

$$\sigma:(0,\infty) \times \Gamma \to P_m\Sigma, \quad \sigma = P_m s.$$

The map s is a C^∞ map when viewed as a map of $\mathbb{R}_+ \times \Gamma$ in H. The Jacobian of σ at some point t_0, u_0 is given by

$$(D\sigma)(t_0,u_0) = [-P_m N(S(t_0)u_0),\ P_m(S'(t_0,u_0)]$$

where $N(u) = \varepsilon A^2 u + Au + B(u,u) - f$ and where $S'(t_0,u_0)$ is the application that assigns to v_0 the value $v(t_0)$ of the solution of (15.37), (15.38), the linearized equation along $S(t)u_0$. Now assume that for some $t_0 > 0$, $u_0 \in \Gamma$, $(D\sigma)(t_0,u_0)$ would not be invertible. Then there would exist a tangent vector to Γ, $v_0 \in P_m H$ and a real number α such that the vector $w(t) = \alpha N(S(t)u_0) + S'(t,u_0)v_0$ satisfies $P_m w(t_0) = 0$. Now let $Q(t)$ be the projector on the linear space $N(S(t)u_0)\mathbb{R} + S'(t,u_0)(T_{u_0}(\Gamma))$. Observe $Q(t)w(t) = w(t)$. Also $Q(t)$ solves the transport equations (15.13), (15.14) with $Q_0 = Q(u_0)$ the projector on the space $\mathbb{R}\,N(u_0) + T_{u_0}(\Gamma)$. The choice of Γ (property (I)) implies $\Lambda(Q_0) < \frac{\lambda_m + \lambda_{m+1}}{2}$. From the spectral blocking property (Theorem 15.3) $\Lambda(Q(t)) \le \frac{\lambda_m + \lambda_{m+1}}{2}$. On the other hand, since $Q(t_0)w(t_0) = w(t_0)$ we get by the definition of $\Lambda(Q(t))$ that

$$\frac{\|w(t_0)\|^2}{|w(t_0)|^2} \le \Lambda(Q(t)) \le \frac{\lambda_m + \lambda_{m+1}}{2}$$

Now at t_0, $P_m w(t_0)$ was assumed to be zero. Thus $\|w(t_0)\|^2 \ge \lambda_{m+1}|w(t_0)|^2$ and we arrive at the contradiction $\lambda_{m+1} \le \frac{\lambda_m + \lambda_{m+1}}{2}$. Thus we proved

(15.80) $\quad \sigma$ is regular at each $(t,u_0) \in (0,\infty) \times \Gamma$.

It follows that σ is locally invertible. The local inverses of σ form an atlas for Σ. Thus Σ is a smooth manifold. Also, clearly $P_m\Sigma$ is open in $P_m H$. Now, from property (IV) and Proposition 15.7, $\Sigma \subset C_{m,X}$. Moreover, $\Sigma \cap X = \emptyset$. Indeed, if $x \in \Sigma \cap X$ then $x = S(t)u_0$ with $u_0 \in \Gamma$. But since $S(t)X = X$ it would follow that $x = S(t)y$ with $y \in X$ and, from the injectivity of $S(t)$ that $u_0 \in \Gamma \cap X$. Now this is absurd because Γ was taken to be far away from X. But $\Sigma \subset C_{m,X}$ and $\Sigma \cap X = \emptyset$ imply $P_m\Sigma \cap P_m X = \emptyset$ (just use the definition of $C_{m,X}$). Now let $p \in \overline{P_m\Sigma}$ = the closure in H_m of $P_m\Sigma$. Thus $p = \lim_k P_m S(t_k)u_k$ with $t_k \in (0,\infty)$, $u_k \in \Gamma$. Assuming that u_k

converges to u it follows that p is either in P_mX (if t_k have ∞ as cluster point), or in Γ (if 0 is a cluster point of t_k) or in $P_m\Sigma$ (if a finite nonzero number is a cluster point for the t_k's.) So $\overline{P_m\Sigma} \subset P_mX \cup P_m\Sigma \cup \Gamma$ and the union is disjoint. Now we claim that

$$\mathcal{E}_m(R_0) = \{u \in P_mH \mid \|u_m\| \leq R_0\} \subset \overline{P_m\Sigma} .$$

We argue by contradiction and assume that a small open ball $\{p \in P_mH \mid |p - p_0| < \varepsilon\} = B_\varepsilon$ is contained in $\|p_m\| < R_0$ but does not intersect $P_m\Sigma$. It follows that $B_\varepsilon \cap P_mS(t)\Gamma = \emptyset$ for all $t > 0$. Now, from the isoperimetric inequality

$$\mathrm{vol}_{m-1}(\partial B_\varepsilon) \leq \mathrm{vol}_{m-1}(P_m S(t)\Gamma) \leq \mathrm{vol}_{m-1} S(t)\Gamma .$$

Now the right-hand side decreases exponentially as $t \to \infty$ and the left-hand side is a positive number, absurd. (Here we used the fact that m-1 dimensional volume elements decay exponentially, that Γ is compact and that P_m, as an orthogonal projector, does not increase volumes). Now Σ is arcwise connected, obviously. Thus $P_m\Sigma$ is connected,open. Thus, since σ is regular, Σ is a covering space of $P_m\Sigma$. Therefore the cardinal of $P_m^{-1}\{p\} \cap \Sigma$ is equal to the number of connected components of Σ, that is, it is equal to one. Thus $P_m:\Sigma \to P_m\Sigma$ is injective. Now we define the map ϕ by $\phi:\mathcal{E}_m(R_0) \to H$

$$\phi(p) = \begin{cases} p & \text{if } p \in \Gamma \\ u \in \Sigma & \text{if } p \in P_m\Sigma,\ P_mu = p \\ u \in X & \text{if } p \in P_mX,\ P_mu = p \end{cases}$$

The map ϕ is clearly well defined. Now we will show that ϕ is Lipschitz. Let $p_1, p_2 \in \mathcal{E}_m(R_0)$. Let us take the straight line that joins them,

$\rho(\tau) = p_1 + \tau(p_2 - p_1)$, $\tau \in [0,1]$. Assume first that $p(\tau) \notin P_m X$ for all $\tau \in (0,1)$. (Since $d_M X = m-1$ this is the case for most points.) Then, for each $\tau \in [0,1]$, $p(\tau) = P_m u(\tau)$, with $u(\tau)$ Σ, $u(\tau)$ a smooth curve. Now $u(\tau) = S(t)u_0$ for some $t > 0$, $u_0 \in \Gamma$ depending on τ. It follows that the tangent $du/d\tau$ to $u(\tau)$ is the transported $S'(t,u_0)v$ for some $v \in T_{u_0}(\Gamma) + N(u_0)\mathbb{R}$. Since this space is in $C_{m,1/3}$ by assumption (V) and since $C_{m,1/3}$ is invariant to $S'(t,u_0)$ by Theorem 15.5 it follows that $du/d\tau \in C_{m,1/3}$. Therefore,

$$|(1 - P_m)(u(1) - u(0))| \leq \int_0^1 |(1 - P_m) \frac{du}{d\tau}| d\tau$$

$$\leq \frac{1}{3} \int_0^1 |P_m \frac{du}{d\tau}(\tau)| d\tau = \frac{1}{3} \int_0^1 |p_2 - p_1| d\tau = \frac{1}{2} |p_2 - p_1|.$$

Thus

$$(15.81) \qquad |\Phi(p_1) - \Phi(p_2)| \leq \frac{4}{3} |p_1 - p_2|.$$

On the other hand, if $p(\tau) \in P_m X$ for some $\tau \in (0,1)$, say $p(\tau) = P_m u_\infty$, then since $\bar{\Sigma} \subset C_{m,X}$, it follows $\phi(p_1) - u_\infty \in C_{m,1/3}$ and $\phi(p_2) - u_\infty \in C_{m,1/3}$. Thus

$$|(1-P_m)(\Phi(p_2) - \Phi(p_2)| \leq |(1-P_m)(\Phi(p_2) - u_\infty)| + |(1-P_m)(\Phi(p_1) - u_\infty)|$$

$$\leq \frac{1}{3} |P_m(\Phi(p_2)) - u_\infty)| + \frac{1}{3} |P_m(\Phi(p_1) - u_\infty)|$$

$$= \frac{1}{3} |p_2 - p(\tau)| + \frac{1}{3} |p_1 - p(\tau)| = \frac{1}{3} |p_2 - p_1|.$$

The last equalilty is true because p_1, $p(\tau)$, p_2 are on a straight line with $\rho(\tau)$ between p_1 and p_2. Thus (15.81) is valid for every p_1, p_2.

Finally, let us consider the problem of uniform exponential convergence of trajectories to $\bar{\Sigma}$. Let $u_0 \in H$ be arbitrary and let us consider

$S(t)u_0$. If $S(t)u_0$ stays outside $C_{m,X}$ for all $t \geq 0$ then, according to Proposition (15.7)

$$\text{dist}(S(t)u_0, X) \leq c \exp\left(- \frac{\lambda_m + \lambda_{m+1}}{4} t\right)$$

with $c = \max_{x \in X} |x - u_0|$. Since $X \subset \bar{\Sigma}$, (this follows from $\overline{P_m\Sigma} \supset P_m X$ and the fact that $\bar{\Sigma} \subset C_{m,1/3}$, thus $P_m u = P_m x$, $u \in \bar{\Sigma}$, $x \in X$ implies $u = x$), clearly exponential decay towards X implies exponential decay towards $\bar{\Sigma}$. We may assume therefore that $S(t)u_0 \in C_{m,X}$ for $t \geq t_0$ (again from Proposition (15.7)). If we take t large enough then $P_m S(t)u_0$ belongs to $\mathcal{E}_m(R_0)$, clearly. (See (15.8)). It follows that $P_m S(t)u_0$ belongs to $P_m \Sigma$. Indeed the other two possibilities, namely $P_m S(t)u_0$ in $P_m X$ and $P_m S(t)u_0$ in Γ are excluded easily. If $P_m S(t)u_0 \in P_m X$, since $S(t)u_0 \in C_{m,X}$, it follows that $S(t)u_0 \in X$ and the distance to $\bar{\Sigma}$ (and X) are 0. If t is large enough $\| P_m S(t)u_0 \| < R_0$ so $P_m S(t)u_0 \notin \Gamma$. Now, since $P_m S(t)u_0 \in P_m \Sigma$, there exists $u_1 \in \Sigma$ such that $P_m(S(t)u_0) = P_m(u_1)$. We claim that, for each $x \in X$,

$$|u_1 - S(t)u_0| < |S(t)u_0 - x|.$$

Indeed, since $S(t)u_0 \in C_{m,X}$ we have

$$|(1 - P_m)(S(t)u_0 - x)| \leq \frac{1}{3} |P_m(S(t)u_0 - x)| = \frac{1}{3} |P_m(u_1 - x)|$$

and also since $\Sigma \subset C_{m,X}$

$$|(1 - P_m)(u_1 - x)| \leq \frac{1}{3} |P_m(u_1 - x)|.$$

But $S(t)u_0 - u_1 = (1 - P_m)(S(t)u_0 - u_1)$ and thus

$$|S(t)u_0 - u_1| = |(1 - P_m)(S(t)u_0 - u_1)|$$

$$\leq \frac{2}{3} |P_m(u_1 - x(| \leq \frac{2}{3} |u_1 - x| < |u_1 - x|$$

since $\Sigma \cap X = \emptyset$. Thus, for each fixed $S(t)u_0$ for t sufficiently large, the distance between $S(t)u_0$ and $\bar{\Sigma}$ must be attained on Σ. Let us fix t and consider $u = S(t)u_0$ and $u_1 \in \Sigma$ such that $\mathrm{dist}(u;\bar{\Sigma}) = |u - u_1|$. Let $Q(u_1)$ be the projector on the tangent space at u_1 to Σ. Clearly

$$Q(u_1)(u - u_1) = 0.$$

On the other hand, since Σ is an integral surface

$$\lambda(Q(u_1)) \geq \frac{\lambda_m + \lambda_{m+1}}{2}$$

because of property (II) and the spectral blocking property. From the definition of $\lambda(Q(u_1))$ it follows that

$$\frac{\|u - u_1\|^2}{|u - u_1|^2} \geq \frac{\lambda_m + \lambda_{m+1}}{2}.$$

Let us consider $u(s) = S(s)u$, $u_1(s) = S(s)u_1$. Clearly, $u(s) = S(s + t)u_0$ and $u_1(s) \in \Sigma$. Forming the difference $v(s) = u(s) - u_1(s)$ and taking the first energy estimate we get

$$\frac{1}{2}\frac{d}{ds}|v(s)|^2 + \|v\|^2 \leq cR_0|v|\|v\| \leq \frac{1}{2}\|v\|^2 + cR_0^2|v|^2.$$

Thus

$$\frac{d}{ds}|v(s)|^2 \leq -\|v\|^2 + cR_0^2|v|^2 = -|v|^2\left[\frac{\|v\|^2}{|v|^2} - cR_0^2\right].$$

Computing at $s = 0$ we get

$$\frac{d}{ds}|v(s)|^2\Big|_{s=0} \leq -|v|^2\left[\frac{\lambda_m + \lambda_{m+1}}{2} - cR_0^2\right] < 0$$

Since

$$\frac{d}{ds}|\mathrm{dist}(S(t+s)u_0, \bar{\Sigma})|^2\Big|_{s=0} = \frac{d}{ds}|v(s)|^2\Big|_{s=0}$$

we obtain the desired uniform exponential decay

$$\text{dist}(S(t)u_0, \overline{\Sigma}) \leq c \exp\left(-\frac{\lambda_m + \lambda_{m+1}}{4} t\right)$$

for $t \geq t_0$.

We proved therefore

Theorem 15.9. Let $R_0 = 4\|f\|$. Let m be such that $\lambda_{m+1} > \lambda_m$ and $\lambda_m \geq c\ (\|f\|)$ where $c_\varepsilon\ (\|f\|)$ is a positive constant depending on $\|f\|$. Let

$$\mathcal{E}_m(R_0) = \{u|\ \|u\| \leq R_0\} \cap P_mH$$

Let $\Gamma = \{u|\ \|u\| = R_0\} \cap P_mH$ and $\Sigma = \bigcup_{t>0} S(t)\Gamma$. Then $\overline{\Sigma}$ is an inertial manifold for the equation (15.3). Moreover, Φ: $\mathcal{E}_m(R_0) \to \overline{\Sigma}$ defined by $\Phi(p) = u$ if $u \in \overline{\Sigma}$, $P_mu = p$, is Lipschitz

$$|(1 - P_m)(\Phi(p_1) - \Phi(p_2))| \leq \frac{1}{3}|p_1 - p_2|.$$

Also Σ is a C^∞ manifold and the projectors Q(u) onto the tangent space to Σ at u satisfy

$$\lambda_m \leq \Lambda(Q(u)) \leq \frac{\lambda_m + \lambda_{m+1}}{2} \leq \lambda(Q(u)) \leq \lambda_{m+1}.$$

For any $u_0 \in H$, there exists t_0 and c depending on $|u_0|$ only such that

$$\text{dist}(S(t)u_0, \overline{\Sigma}) \leq c \exp\left(-\frac{\lambda_{m+1} + \lambda_m}{4} t\right)$$

for $t \geq t_0$.

One can prove also

Theorem 15.10. The inertial manifold $\overline{\Sigma}$ has the following asymptotic completeness property. For each $u_0 \in H$, there exists t_1 and $u_1 \in \Sigma$ such that

$$\lim_{t \to \infty} |S(t)u_0 - S(t - t_1)u_1| = 0$$

BIBLIOGRAPHY

The material in chapters 1 and 2 can be found in [A1], [T1], [LM1]. For chapter 3 we used [K1] and [G]. Most of chapters 4, 5, and 6 are well known ([LM1], [T1], [T2], [Be-Lo]). In chapters 7, 8, 9, 10, and 12 we present the classical theory of Navier Stokes initiated by Leray ([La1], [Li1], [T1], [T2]). Chapter 11 illustrates the results of [K1] in the periodic case. Chapters 13 and 14 cover more recent results, ([CF1], [CFT1], [CFT2], [T3]). Work in progress is presented in chapter 15 ([CFNT1], [CFNT2] [F-S-T]).

[A1] S. Agmon - Lectures on elliptic boundary value problems, Van Nostrand Math. Studies, no. 2, 1965.

[ADN] S. Agmon, A. Douglis, L. Nirenberg - Estimates near the boundary for solutions of elliptic partial differential equations satisfying general boundary conditions II, Comm. Pure Appl. Math 17 (1964), 35-92.

[Be-Lö] J. Bergh, J. Löfström - Interpolation spaces, An introduction. Grundlehren der Mathematischen Wissenschaften 223, Springer, Berlin, Heidelberg, NY, 1976.

[Ca1] L. Cattabriga - Su un problema al contorno relativo al sistema di equazioni di Stokes, Rend. Sem. Mat. Univ. Padova 31, (1961), 308-340.

[C1] P. Constantin - Note on loss of regularity for solutions of the 3-D incompressible Euler and related equations, Comm. Math. Phys. 104, (1986), 311-32.

[C2] P. Constantin - Collective L^∞ estimates for families of functions with orthonormal derivatives, to appear in Indiana Journal of Math.

[CF1] P. Constantin, C. Foias - Global Lyapunov exponents, Kaplan-Yorke formulas and the dimension of the attractors for 2-D Navier-Stokes equations, Comm. Pure and Applied Math. 38 (1985), 1-27.

[CFT1] P. Constantin, C. Foias, R. Temam - Attractors representing turbulent flows, Memoir of AMS, January, 1985, Volume 53, No. 314.

[CFT2] P. Constantin, C. Foias, R. Temam - On the dimension of the attractors in two dimensional turbulence, preprint.

[CFNT1] P. Constantin, C. Foias, B. Nicolaenko, R. Temam - Integral manifolds and inertial manifolds for dissipative partial differential equations, preprint.

[CFNT2] P. Constantin, C. Foias, B. Nicolaenko, R. Temam - Comptes Rendus de l'Acad. Sci. Paris, vol. 302, série I, no. 10, 14 Mars 1986, 1375-378.

[F-S-T] C. Foias, G. R. Sell, R. Temam - Comptes Rendus de l'Acad. Sci. Paris 201, série I, 1985, 139-141.

[G] J.-M. Ghidaglia - Régularité des solutions de certaines problèmes aux limites linéaires liées aux équations d'Euler, Comm. P.D.E., vol. 9, No. 13, (1984), 1264-9.

[K] R. H. Kraichnan - Inertial ranges in two dimensional turbulence, Phys. Fluids, 10, (1967), 1417-1423.

[K1] T. Kato - Nonstationary flows of viscous and ideal fluids in R^3, J. Funct. Anal. 9, (1972), 296-305.

[Ko1] A.N. Kozhevnikov - On the operator of the linearized steady state Navier-Stokes problem, Math. USSR Sbornik, vol 53, No.1, (1986), 1-16.

[La] O. A. Ladyzhenskaia - The mathematical theory of viscous incompressible flow, Gordon and Breach, 1969.

[L1] E. H. Lieb - An L^p bound for the Riesz and Bessel potentials of orthonormal functions, J. Funct. Anal, 51, No.2, (1983), 159-165.

[L-L] L. Landau, I. M. Lifschitz - Fluid Mechanics, Addison-Wesley, New York, 1953.

[L-T] E. H. Lieb, W. Thirring - Inequalities for the moments of the eigenvalues of the Schrödinger equation and their relation to Sobolev spaces, in Studies in Mathematical Physics: Essays in honor of Valentin BarYman, (E. H. Lieb, B. Simon and A. Wightman, eds.) Princeton Univ. Press, Princeton, NJ, 1976.

[Li1] J. L. Lions - Problèmes aux limites dans les équations aux derivées partielles, Presses de l'Université de Montreal, Montreal 1965.

[LM1] J. L. Lions, E. Magenes - Problèmes aux limites non homogenes et applications, Dunod, Paris 1968-1970.

[R1] I. Richards - On the gaps between numbers which are sums of two squares, Adv. In Math., 46 (1982), 1-2.

[So1] V. A. Solonnikov, On estimates of Green tensors for certain boundary problems, Dokl. Akad Nauk SSSR, 29 (1960), 988-991.

[T1] R. Temam - Navier-Stokes equations: theory and numerical analysis, North Holland, Amsterdam, New York 1977.

[T2] R. Temam - Navier-Stokes equations and nonlinear functional analysis, SIAM, Philadelphia 1983.

[T3] R. Temam - Infinite dimensional dynamical systems, Proc. Symp. Pure Math., vol. 4J, part 2, (1986), 431-445.

[V-Y1] I.I. Vorovich, V. I. Yudovich, Dokl Akad. Nauk SSSR, 124 (1959), 542.

INDEX